THE FLORA OF

MOTUPORE ISLAND

PAPUA NEW GUINEA

HELEN FORTUNE HOPKINS

AND

JAMES I. MENZIES

UNIVERSITY OF PAPUA NEW GUINEA PRESS

1995

Front cover; *Cupaniopsis curvidens*

The Flora of Motupore Island, Papua New Guinea

ISBN 9980-84-055-2

University of Papua New Guinea Press and Bookshop
PO Box 413
University PO, NCD
Papua New Guinea

THE FLORA OF MOTUPORE ISLAND

CONTENTS

Acknowledgements

We are grateful to the former chairman of the Motupore Island Research Department, Dr. Mike Huber, and the staff at Motupore including Terry Frohm, John Rewald, Sai Gwae and the late John Harry Hilary for their help and cooperation; Kipling Naoni and Max Kuduk for help with curation of herbarium specimens; Dick Knight for the Loloata rainfall data; Drs. Greg Leach and David Frodin for critical reviews of a preliminary manuscript and Kwani Kego for secretarial work. Particular thanks are due to David Frodin for allowing free access to his unpublished notes and checklists. Jones Hiaso drew Figure 1 and John Sengo drew Figure 4.

Finally, we thank the Papua New Guinea Biological Foundation for funding publication of this work.

1. INTRODUCTION

The aim of this book is to help students and others identify the terrestrial vascular plants on the small island of Motupore in Bootless Bay, Central Province, on the south coast of Papua New Guinea (Fig. 1). We hope it will encourage further botanical and ecological studies on the island.

Since the plants on Motupore are found elsewhere in the surroundings of Port Moresby, the book may be of some use as a general guide to the local flora. However, the flora of Motupore is much less diverse than that of similar vegetation types on the mainland, and a number of common trees are absent from the island.

The main part of the text is an annotated and illustrated checklist of species, arranged alphabetically by family. Although mangroves and grasses are included, details of these are available elsewhere. The book is not intended as a critical flora, and names have been taken primarily from the checklists and herbarium specimens at the University of Papua New Guinea (UPNG), and the taxonomic literature available there. Some identifications have been checked at the Royal Botanic Gardens at Kew, and a few specimens have been seen by specialists (Asclepiadaceae and Apocynaceae - Paul Forster; Cyperaceae - Paul Goehts; Gramineae - J. Veldkamp; Oleaceae - R. Kiew; Rubiaceae - S. Sohmer & O. Gideon; Vitaceae - B. Jackes). However, errors and omissions are inevitable and corrections are welcomed.

The notes on vegetation types and on the distribution of species on the island were written prior to a major fire in October, 1992. Some notes on the effects of the fire have been added, but the species accounts have not been rewritten.

The first plant specimens from Motupore preserved in the herbarium at the University of Papua New Guinea were collected in 1972 by David Frodin and Jocelyn Powell, and added to in 1975 by Camilla Huxley. Subsequent collections were mostly by other staff and students of UPNG, especially Frodin. The first checklist was compiled in 1976-77 and subsequently updated several times by Frodin. After Frodin's departure in 1985, his checklists were updated and computerised by M. Hopkins, L. Balun and P. Piskaut.

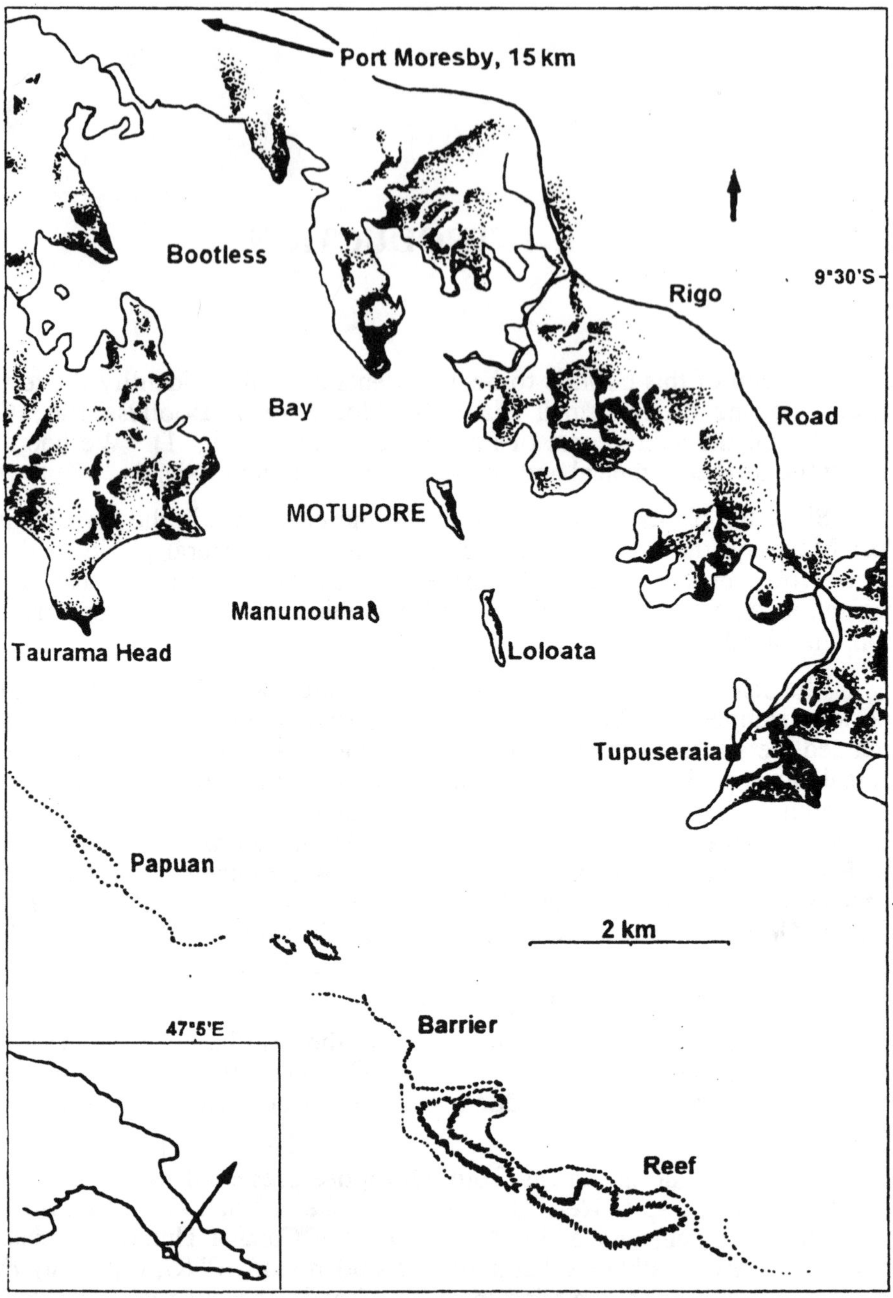

Figure 1. Papua New Guinea, Bootless Bay and Motupore Island.

2. MOTUPORE ISLAND

Motupore is one of three small, steep-sided islands in Bootless Bay about 15 km SE of Port Moresby (Fig. 1); the others are Loloata and Lion Island or Manunouha. Motupore is about 1 km from the mainland, measures approximately 800 x 200 m, has an area of about 13 ha and reaches an elevation of 60 m (Barratt, 1972). The axis of the island runs NW-SE and the ridge is fairly level, but slopes steeply on either side to the sea. At the north end is a flat sandy platform, about 1 m above sea level, where the buildings, including the marine research station, are situated (Fig. 2), and there is a curved sandy spit, exposed only at low water, at the north east corner.

Geologically the island is composed of bands of calcareous shale and chert laid down as deep marine sediments in the Eocene, and now tightly folded and faulted, and dipping steeply at almost 90° (H. Davies, pers. comm.). These Eocene deposits, called the Port Moresby Beds, extend in a narrow coastal strip as far south as Hood Lagoon. The island of Motupore was formed from a ridge crest when the area subsided and was then covered by the sea. Loloata was part of the same ridge crest but the two islands are now separated by a fault (H. Davies, pers. comm.). The soil is a red gravelly clay common throughout the surroundings of Port Moresby (Barratt, 1972).

The annual rainfall for the region is about 1000-1200 mm (Heyligers, 1982) but highly variable from year to year. The climate is strongly seasonal with most rain usually falling between November and April, and very little during the rest of the year (Fig. 3). Although the pattern is much the same from year to year, the exact timing and amount of the rains vary. The SE Trade Winds (Laurabada) blow almost daily from May to October, though they may start as soon as March and may continue till the end of November or even into December and they rarely bring rain. After a period of doldrums, the NW Monsoon Winds (Lahara) then bring intermittent rain, heaviest in January-March. They blow less steadily than the Trades and are interspersed with doldrums and light breezes. When the Lahara winds blow, Bootless Bay is more or less sheltered by promontaries on the mainland, towards Port Moresby. Thus its rainfall pattern is not identical to that for the city (Moore, 1982). Moreover, records from the Motupore weather station over the period 1992-4 show considerable variation even from Loloata which is less than 1 km distant. There is no surface water on the island.

Currents around Motupore are affected by the tides and winds. Motupore lies within a lagoon formed by the offshore Papuan Barrier Reef, and the currents tend to be channelled inside this lagoon, especially when the SE Trades are blowing (Moore, 1982).

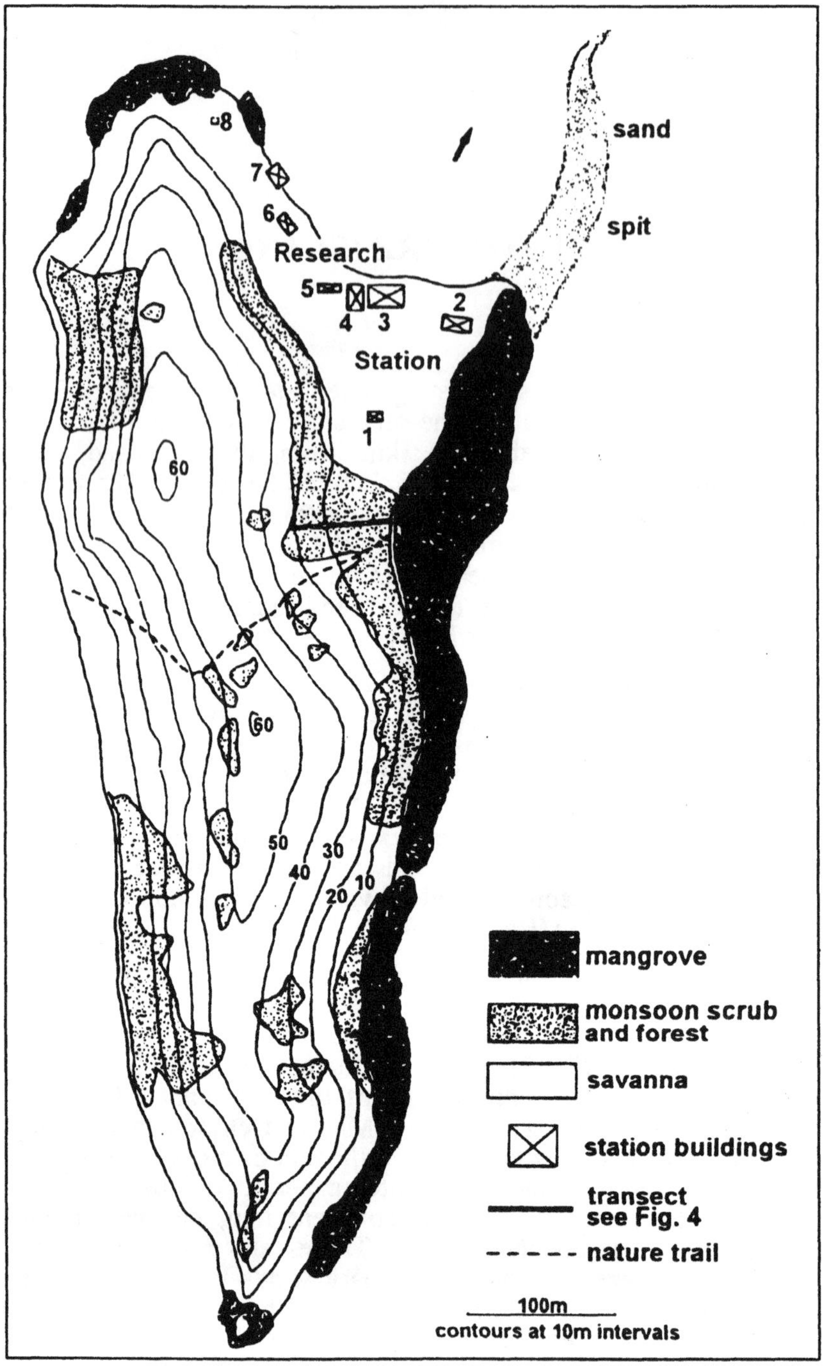

Figure 2. Motupore Island.

Key to station buildings: 1=Caretaker's house; 2=new hostel; 3=old hostel; 4=dry lab; 5=wet lab; 6=house 2; 7=house 1; 8=generator.

Motupore Island was acquired by the University of Papua New Guinea in 1969, primarily as an archaeological site (Barratt, 1972), as the flat sandy area at the north end of the island had been occupied by a prehistoric Motu settlement. Dating of midden material indicates that the island was occupied from about 1200 to 1700 AD (Worthing, 1978; Swadling, 1981). Oral tradition relates that fighting with the people from Tupusereia led to the abandonment of the settlement on Motupore in favour of Taurama (Swadling, 1981). Excavations on the island ceased by the mid-1980s and its use as a marine station started in 1974 though a wet lab was not built until 1976. The island is now used for research and student field courses in geography, geology, environmental science and biology.

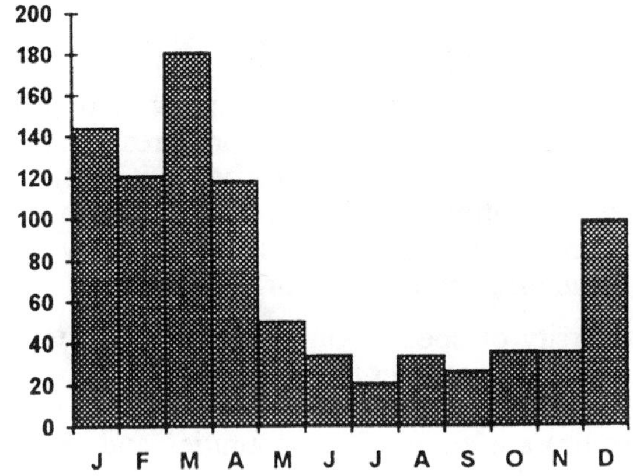

Figure 3. Mean monthly rainfall (mm) on Loloata Island based on 12 years from 1979-1991.

A nature trail was laid out in the island in the 1970s and modified a few years later (Fig. 2). The first guide to the trail by Frodin & Johnstone (no date) contained botanical information only while the second (Brooks *et al.*, no date) also included much useful background information on geology, soils, etc.

A bibliography of scientific papers relating to Motupore was published by McGregor & Huber in 1993.

3. PLANT COMMUNITIES

From a distance, as it is approached by boat or seen from the air, Motupore appears as a relatively well-wooded island with some patches of open grassland on the sides of the ridge and a fringe of dark green mangroves on the eastern, landward side. Most of the island is covered with eucalypt woodland but there are patches of monsoonal scrub or forest, especially towards the bottom of the slopes. Where this reaches the beach it merges with the trees and shrubs of the strand vegetation. At the north end of the island, where the marine research station is situated, there is a flat grassy area with numerous introduced plants, including coconuts and ornamentals.

The overall diversity of species on Motupore is much lower than on the mainland of the southeast Papuan region but the same communities are present. Two of these (mangrove and strand) are influenced by the proximity of the sea, two (savanna and scrub) are climatic communities and the fifth (in the station area) is anthropogenic. All the vegetation types are strongly influenced by the seasonal climate.

A. Mangrove

Mangrove is a vegetation type consisting of trees or, less commonly, shrubs growing on coasts between high and low tide levels. It presents a rather uniform appearance and is evergreen; species of Rhizophoraceae have dark, glossy green foliage while the leaves of *Avicennia* species are paler and rather glaucous. Factors influencing mangrove vegetation include water-logged, oxygen-deficient substrate, salinity, and tides and this results in the zonation of species and numerous structural adaptations to their specialised environment (e.g. salt glands, stilt roots, pneumatophores, floating propagules and viviparity) (see Frodin *et al.*, 1982; Tomlinson, 1986).

The mangrove on Motupore includes about 11 species in ten genera, out of a total of about 25 species in 12 genera (Frodin *et al.*, 1982) recorded in the Port Moresby region. *Excoecaria agallocha*, which is sometimes considered a mangrove, is excluded because on Motupore it is really part of the strand flora. Other tree and shrub species such as *Erythrina variegata* and *Pongamia pinnata* may grow within the upper limit of the mangrove but are not considered to be mangrove species.

The principal mangrove species on Motupore are listed in Table 1.

A few other mangrove species have been recorded intermittently including *Heritiera littoralis* (Sterculiaceae), *Xylocarpus granatum* (Meliaceae), and *Scyphiphora hydrophyllacea* Gaert. (Rubiaceae) (Frodin, no date) but do not seem to have become established.

The best developed area of mangrove on Motupore forms a narrow band down the sheltered, landward side, widest at the northern end (Plate 1), and dominated by *Rhizophora* species whose stilt roots make much of it impenetrable. There is also some *Bruguiera* and *Ceriops*, with occasional *Sonneratia*, and a patch of *Aegiceras* about half way along. *Osbornia* is common as a low bush on the landward side; *Lumnitzera* is found at the north end only, near the New Hostel but there is little *Avicennia*.

Table 1. Mangrove species currently found on Motupore Island

Family	Species
Combretaceae	*Lumnitzera racemosa*
Myrsinaceae	*Aegiceras corniculatum*
Myrtaceae	*Osbornia octodonta*
Plumbaginaceae	*Aegialitis annulata*
Rhizophoraceae	*Bruguiera gymnorrhiza*
	Ceriops tagal var. *tagal*
	?Rhizophora apiculata
	Rhizophora lamarckii
	Rhizophora stylosa
Sonneratiaceae	*Sonneratia alba*
Verbenaceae	*Avicennia eucalyptifolia*

The western side of the island is more exposed and along the bare, rocky beach there is only a narrow and intermittent band of trees, not a true mangrove formation; *Aegialites annulata* is the commonest species and reaches c. 1 m but there are a few specimens of *Avicennia* and *Rhizophora* (Plate 2).

B. Strand vegetation

This refers to the narrow fringe of plants growing just above the high water mark around the edge of the island and includes a number of distinctive species. It is semi-deciduous, and must be relatively tolerant of salt. It is a specialised but geographically widespread community, due largely to the dispersal of fruits and seeds by ocean currents.

Where the mangrove is well developed there is usually a bare rocky or sandy strip separating it from the strand flora behind (Plate 3). In many places the strand vegetation merges with scrub on the hill side.

The strand is dominated by trees such as *Cordia subcordata, Thespesia populnea* and *Guettarda speciosa*. Less common trees include *Excoecaria agallocha, Pongamia pinnata* and *Xylocarpus rumphii*, while shrubs such as *Clerodendrum inerme, Premna serratifolia, Maytenus emarginata, Colubrina asiatica* and *Vitex trifolia* may be locally common. Some strand species that are generally common in the Indo-Pacific are only occasional or rare (*Sophora tomentosa, Hibiscus tiliaceous, Terminalia catappa*) and others are absent e.g. *Calophyllum inophyllum* L. (Guttiferae), *Barringtonia asiatica* L. (Lecythidaceae) and *Casuarina equisetifolia* Forst. (Casuarinaceae) (except for a few planted at the north end of island). *Scaevola sericea* has been recorded but is not apparently still present.

At the northeast corner near the New Hostel is a small patch of typical beach vegetation on sandy ground, dominated by *Ipomoea pes-capri* and *Sesuvium portulacastrum*.

C. Eucalyptus savanna

This is an open woodland dominated by *Eucalyptus* trees with a ground cover of grasses (Plate 4), often referred to as *Eucalyptus* savanna, though the term 'woodland' is preferred by some authors. The density of trees varies so that in some places there is a broken canopy, and elsewhere there are open patches of *Themeda* or other grasses. This type of vegetation covers 60-70% of Motupore (Menzies & Sengo, 1993).

The savanna is much affected by the seasons. In the dry season *Eucalyptus* and other trees are partly deciduous and the proportion of leaves dropped depends on the severity of the drought. During the wet season, a number of herbaceous plants appear, e.g. *Mukia maderaspatana*, *Anisomeles malabarica* and *Commelina ensifolia*, some from woody rootstocks and others from seed. The importance of fire in the ecology of *Eucalyptus* woodland is discussed later.

Most of the *Eucalyptus* trees resemble *E. confertiflora*, but some show characteristics of *E. papuana* and this is discussed in the species accounts.

Eucalyptus trees grow up to 10 m and often have crooked trunks with exuding gum and exfoliating bark. Other woody species common in the savanna include the trees *Acacia auriculiformis* and *Timonius timon* and the shrubs *Ficus opposita*, *Canthium suborbiculare*, *Allophylus cobbe*, *Alectryon repandodentatus*, *Clerodendrum floribundum* var. *ovatum*, *Premna dallachyana*, *Breynia cernua* and *Flueggea virosa*. The ground layer is dominated by *Themeda* species, which grow in tussocks to about 1.5 m tall, and other common grasses include *Heteropogon contortus* and *Rhynchelytrum repens*. There are numerous small leguminous creepers in open areas, and larger creepers include *Cissus hastata*, *Secamone elliptica*, *Marsdenia velutina*, *Parsonsia quinquebullata*, *Rhyssopteris timoriensis*, *Canavalia papuana* and *Jasminum* species. On the eastern side especially, there are scattered cycads, (*Cycas campestris*) often covered in the parasitic vine *Cassytha filiformis*. The parasitic ground orchid, *Dipodium punctatum*, which lacks leaves, is occasionally seen in the grassland.

D. Monsoonal scrub and forest.

Several terms have been used for this vegetation type. Heyligers (1982), referring to the Port Moresby area, called it 'semi-deciduous scrub and forest', and Brock (1988) referred to similar vegetation in Northern Australia as 'monsoon forest', including within that various sorts of vine forest and thicket. On Motupore the patches are small and the trees never form a continuous, even canopy. Many labels on herbarium specimens from Motupore use the term 'scrub' to denote these areas since the vegetation is usually composed of short trees, scrambling shrubs and creepers, and we have used this term in the species accounts.

This type of vegetation is strongly affected by the climate. Some species avoid water stress during the dry season by being deciduous though the degree of deciduousness varies between species and between years. *Garuga floribunda*, *Albizia carrii* and *Bombax ceiba* (whose pagoda-shape makes it very striking) are deciduous every year while others, including *Grewia* species and *Micromelum minutum* are completely deciduous only during exceptionally dry years. Other species are evergreen with rather tough, dark green leaves (e.g. *Celtis philippinensis*, *Drypetes australasica*, *Mimusops elengi*).

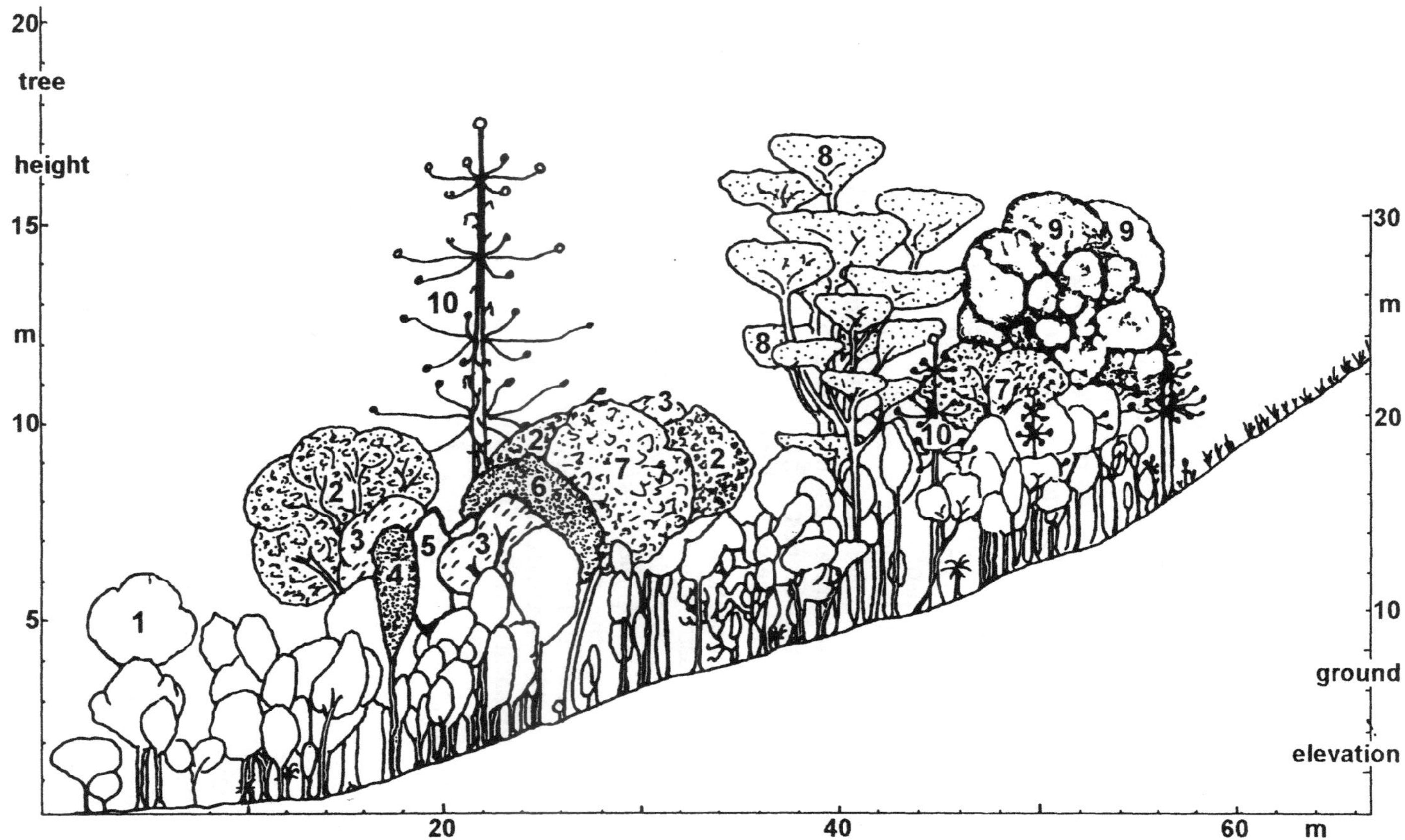

Figure 4. Monsoon forest transect (see Figure 2 for location) from Menzies & Sengo, 1993.

Only trees over 5 m high are labelled. 1=*Bridelia oligantha*; 2=*Canthium suborbiculare*; 3=*Ellatostachys tetraporandra*; 4=Sapindaceae, sp. indet.; 5=*Alectryon repandodentatus*; 6=*Turraea virens*; 7=*Exocarpus latifolius*; 8=*Albizia carrii*; 9=*Elaeocarpus arnhemicus*; 10=*Bombax ceiba*.

There are several isolated patches of this vegetation on the island, especially on the lower slopes, covering about 30% in total. Well developed areas are on the sheltered northeast corner (Plate 5; Fig. 4), near the station and towards the southwest end where scrub extends from strand to ridge top.

In the northeast corner, the canopy is irregular and formed by the crowns of various species including *Albizia carrii*, *Elaeocarpus arnhemicus*, *Garuga floribunda*, *Mimusops elengi*, several Sapindaceae and *Bombax ceiba* which is emergent; *Eucalyptus* is absent. *Bombax ceiba* and *Albizia carrii* are the tallest species and reach 15-20 m in height. On the west side, *Bombax* is absent and, where a canopy is formed, *Garuga* and *Albizia* are the common trees but a few large *Ficus microcarpa* are present at the southern end. Elsewhere on the island the canopy of the scrub patches is lower, less than 10 m high. Widespread woody species in this sort of vegetation include *Bridelia oligantha*, *Drypetes australasica*, *Harpullia leptococca*, *Elattostachys tetraporanda*, *Exocarpus latifolius*, *Celtis philippinensis*, *Litsea glutinosa* and species of Rutaceae.

Tree crowns are usually swathed in woody creepers, especially *Jasminum* spp. and various species of Asclepiadaceae and Vitaceae and there are numerous straggly shrubs, also known as shrub-lianes (Heyligers, 1982). Several of the creepers have spines or thorns (e.g. *Capparis quiniflora* and *C. sepiaria*, *Harrisonia brownii*, *Maclura cochinchinensis*, *Smilax* sp.). Most patches of monsoonal thicket are almost impenetrable and some are inaccessible because they are on very steep slopes. There is almost no herb layer because of the density of the canopy, and the ground is carpeted in dead leaves.

Towards the shore line, the monsoonal scrub often merges with the strand vegetation to form a coastal thicket, with species typical of both vegetation types entwined. These include *Grewia acuminata*, *Capparis* species, *Flagellaria indica* and *Caesalpinia bonduc*. In contrast, there is usually quite a sharp boundary between the monsoonal forest and the savanna (Plate 6). Species frequently found at the boundary include *Turraea virens*, *Alectryon repandodentatus* and *Alstonia spectabilis*, the last particularly in the northeast near the foot of the nature trail. The presence of these three species may indicate a transition from savanna to scrub. There are many dead cycad stems, often covered in ferns, within the patches of forest.

E. Station area

The vegetation at the north end of the island around the marine station and houses is strongly influenced by humans. The flat sandy area between the buildings is covered with grasses and herbaceous weeds, many of which are not native to New Guinea (Plate 7). A number of trees and shrubs have been planted, especially coconuts (*Cocos nucifera* L., Palmae) and ornamentals including frangipani (*Plumeria*, Apocynaceae), *Hibiscus schizopetalus* (Malvaceae) and introduced legumes (*Bauhinia variegata*, *Adenanthera pavonina*, *Samanea saman*, *Senna siamea* and *Pithecellobium dulce*) near the centre of the area and the custard apple, *Annona squamosa* (Annonaceae) and *Casuarina* (Casuarinaceae) at the northwestern edge.

Some introduced species have become naturalised. These include the small tree *Leucaena leucocephala*, and the garden perennials *Pedilanthus tithymaloides* and *Barleria lupulina*. The cactus *Opuntia stricta* was recorded by Frodin (no date) but had disappeared by 1990.

4. BACKGROUND TO THE FLORA AND VEGETATION

This chapter discuses the composition and relationships of the flora and some of the factors that have influenced the development of the different vegetation types including the dispersal of propagules and the ecology of fire.

A. Composition of the flora

The terrestrial flora of Motupore is composed almost entirely of angiosperms, of which about 55 families are represented. There is one common gymnosperm (*Cycas campestris*), three pteridophytes, one of which is uncommon, few if any bryophytes, and a handful of lichen species. The paucity of ferns and bryophytes is no doubt due to the strongly seasonal climate which results in drought conditions for much of the year, as in the Port Moresby region generally.

Amongst life forms, trees, shrubs and creepers are well represented but epiphytes are scarce. There are several parasitic or hemiparasitic plants notably *Cassytha filiformis*, *Dipodium punctatum*, *Exocarpus latifolius* and *Opilia amentacea*.

Leguminosae is the most significant family in terms of number of species, and includes trees, shrubs, creepers and herbs, many of which have wide distributions, although *Albizia carrii* is a local endemic (see below). Other significant families include Gramineae, Euphorbiaceae and Rubiaceae, and important families of creepers are Asclepiadaceae and Vitaceae. However, the composition of the flora is dynamic and many factors may cause apparent and actual changes. These include the appearance of new species on the strand and in the mangrove (brought by ocean currents) or on land (brought by wind, birds or bats) which may survive for a period and then disappear again. Some herbaceous species are apparently absent in some years and abundant in others for unknown reasons. Fire may also affect the species composition in the ground flora of the savanna.

The following species were noted by Frodin prior to June 1985, but not found since 1990, and they are not mentioned elsewhere in the text: *Glochidion disparipes* Airy Shaw (Euphorbiaceae), *Panicum* sp. and *Sporobolus virginicus* (L.) Kunth (Gramineae), *Cassia 'tora'* L., *Crotalaria chinensis* L., (Leguminosae) and *Abutilon 'auritum'* (Malvaceae).

B. Geographical affinities

The island of New Guinea is included in the botanical region of Malesia, which stretches from Sumatra and the Malay Peninsula in the west as far east as the Louisiade and Bismarck Islands. The overall botanical affinities of Papua

New Guinea are thus with the Indo-Malay Archipelago, and this is particularly true for the rain forest and mountain floras. The vegetation around Port Moresby, however, also has strong affinities with that of northern Australia due to their present geographical proximity, the similarity of climate and the fact that they have been joined by intermittent land bridges over the last 2 million years, and only became separated by the Torres Strait during the mid- to late Pleistocene period (Doutch, 1972).

The mangrove and strand communities contain many widespread genera and species. Some are pantropical (e.g. the genera *Rhizophora* and *Avicennia*; *Ipomoea macrantha*, *I. pes-caprae*, *Caesalpinia bonduc*), others are widespread in the Old World tropics (e.g. the genus *Xylocarpus*; *Vitex trifolia*, *Colubrina asiatica*, *Thespesia populnea*) or from Asia to the Pacific (e.g. the genus *Sonneratia*; *Clerodendrum inerme*, *Sophora tomentosa*, *Guettarda speciosa*).

In the *Eucalyptus* woodland, both *Eucalyptus* and *Acacia* are Australian elements. *Eucalyptus* has its centre of diversity in Australia, and *Acacia* species are very numerous there. The species of the monsoonal forest also have many affinities with Australia and with Malesia too. When Heyligers (1972) analysed the affinities of nearly 300 species from eucalypt savanna and semi-decidous scrub and forest in the Port Moresby area, he found that more than 60% were widespread (either found in Australia and Indo-Malesia (36%) or in Australia, Indo-Malesia and beyond (25%)), 13% were endemic to New Guinea, 13% shared with Australia and 11% shared with the rest of Indo-Malesia.

However, taxonomy can influence our understanding of geographical affinities. If a widespread taxon is uniform from place to place, it will be treated as a single species. If, however, it is rather variable, it may be treated as several local species by some taxonomists, though not necessarily by others. The taxonomy of some species on Motupore, including *Allophylus cobbe*, *Jossinia desmantha*, *Cupaniopsis curvidens*, *Ficus microcarpa* and *Maclura cochinchinensis* is ambiguous and requires further study.

Of particular interest in the Motupore flora are a number of species that are endemic, or nearly so, to the Port Moresby region (Brook *et al.*, no date), namely:

Cycas campestris (Cycadaceae)
Bridelia oligantha (Euphorbiaceae)
Albizia carrii (Leguminosae: Mimosoideae)
Canthium suborbiculare (Rubiaceae)
Jossinia desmantha (Myrtaceae)
Harpullia leptococca (Sapindaceae)

In addition, *Alectryon repandodentatus* (Sapindaceae) is only known from the Port Moresby region and Murray Island in Torres Strait.

C.- Dispersal mechanisms

Dispersal and establishment are important in the ecology of the vegetation of Motupore in two ways. The first concerns the relative distribution of savanna and scrub and the successional change between them (see next section). This is influenced by many factors but succession can only take place if propagules of scrub species are dispersed to, and then become established in savanna areas. The second concerns the diversity of plant communities on Motupore which is lower than in equivalent ones on the mainland. Any increase in diversity on the island requires the dispersal and establishment of propagules from the mainland or other islands.

Four agencies are likely to be significant in the dispersal of propagules to and on Motupore, namely water, wind, birds and bats. In addition, humans may be responsible for the presence of some weeds as well as cultivated plants. A number of species have fruits or seeds that can be dispersed by more than one agency, and this is known as diplochory. For instance, in the case of *Cassytha filiformis*, ocean currents are probably responsible for its wide distribution on tropical beaches since its seeds float, but its occurrence inland may be due to birds (see van der Pijl (1982) who gives a detailed account of dispersal mechanisms in higher plants).

Dispersal by the sea

Many strand and mangrove species have woody or corky fruits or seeds that float and are viable after immersion in salt water. Dispersal by tides and ocean currents explains their wide distribution on tropical coasts, and is one way in which new species can colonise islands.

At certain times of year, large numbers of drift fruits and seeds can be found washed up on the beaches of Motupore. Some belong to species that occur there but others must have come from elsewhere, and these include *Entada* sp., *Barringtonia asiatica* L., *Derris trifoliata* L., *Mucuna gigantea* (Willd.) DC., *Spondias cytherea* Sonn., *Myristica* sp., *Xylocarpus granatum* Ridley, and *Nypa fruticans* (Thunb.) Wurmb.. Many of these propagules germinate readily on the beach (e.g. *Cocos nucifera*), and *Intsia bijuga*, *Heritiera littoralis* and *Xylocarpus granatum* become established intermittently on Motupore from drift seeds. Other types of propagules are sometimes found on the beach, for instance bulbils of *Dioscorea bulbifera* and rhizomes of *Drynaria quercifolia*. Common drift propagules of the New Guinea coast can be identified using Gunn & Dennis (1976) and Peekel (1984).

Dispersal by birds

Many species in savanna and scrub, and a few strand and mangrove ones, have edible fruits or arillate seeds, and birds or bats are likely dispersal agents.

A limited number of bird species has been recorded from Motupore (I. Burrows, pers. comm.; pers. obs.), but includes the following whose diet contains a significant proportion of fruit:

Common Scrub fowl - *Megapodius freycinet*
Bar-shouldered Dove - *Geopelia humeralis*
Emerald Ground Dove - *Chalcophaps indica*
Orange-fronted Fruit-Dove - *Ptilinopus aurantiifrons*
Pied Imperial Pigeon - *Ducula bicolor*
Pheasant Coucal - *Centropus phasianinus*
Helmeted Friarbird - *Philemon buceroides*
Figbird - *Sphecotheres viridis*
Fawn-breasted Bowerbird - *Chlamydera cerviniventris*

Amongst these, fruit-doves and imperial pigeons, especially, are regarded as good dispersal agents. They are exclusively fruit-eaters, they have soft gizzards and do not swallow grit, so they do not grind their food and thus seeds are expelled intact. In addition, they can fly long distances and thus might disperse seeds from other localities to Motupore. Some others, such as bowerbirds, are territorial and so may be more important in local dispersal on the island.

Species likely to be bird-dispersed have fleshy fruits (drupes and berries) or capsules containing arillate seeds. They usually exhibit bright colours and, in the case of capsules, the seeds and fleshy arils are often contrasting. Some examples to illustrate the range of colours shown by plants growing on Motupore are included in Table 2.

Table 2. Fruit colours of some Motupore trees and shrubs.

Fruit colour	Species
pericarp white:	*Flueggea virosa, Tournefortia muelleri*;
pericarp yellow, orange, pink and red:	*Breynia cernua, Drypetes australasica, Exocarpus latifolius, Jossinia desmantha, Mimusops elengi; Maclura cochinchinensis;*
pericarp deep blue:	*Elaeocarpus arnhemicus*;
pericarp shiny black:	*Jasminum aemulum*;
pericarp dull black or blue-black:	*Jasminum didymum, Garuga floribunda*;
capsule and/or aril orange or red, seed black:	*Turraea virens, Harpullia leptococca, Cupaniopsis curvidens, Alectryon repandodentatus;*
capsule red, aril white, seed black:	*Ellatostachys tetraporandra*;
capsule red, aril white, seed brown:	*Maytenus emarginata*;
seed red and black:	*Abrus precatorius*.

Dispersal by bats

Fruit bats (Chiroptera: Pteropodidae), including flying foxes (*Pteropus* species) and their smaller relatives (such as *Syconycteris*, *Rousettus*, *Dobsonia* and *Nyctimene*) are significant seed dispersal agents in many lowland habitats in New Guinea. There are no actual records of fruit bats from Motupore but *Dobsonia moluccensis*, at least, is common in the Port Moresby area and fruit bats of an indeterminate species have been seen at Tahira. The larger species of fruit bats are known to fly long distances in search of food.

Likely bat-dispersed fruits may be green or drab as well as brightly coloured, and fleshy or rather fibrous (van der Pijl, 1957). For large-seeded species, the bat usually ingests just pulp and juice but sometimes large seeds are carried in the mouth while flying and then dropped, or whole fruits may be carried away to a feeding roost. Large-seeded species on Motupore that are bat-dispersed include *Terminalia catappa* and *Guettarda speciosa*. Both these are also found commonly in beach drift and are further examples of diplochory.

Small-seeded fruits are ingested whole and the seeds pass out with the faeces. Likely examples from the Motupore flora include *Ficus*, other Moraceae and *Timonius timon*.

Dispersal by wind

Propagules dispersed by wind include:

a) very small propagules that can be easily blown long distances, e.g. fern spores, orchid seeds.

b) seeds or fruits with hairs attached, e.g. Asclepiadaceae, Apocynaceae and *Bombax ceiba* (all of which have capsular fruits containing numerous plumed seeds); some Compositae (which have single-seeded, plumed fruits).

c) papery, winged fruits e.g. *Rhyssopteris timoriensis*, *Dioscorea* species. These may be able to float on water for short distances also.

The wind blows from the SE for much of the year, that is from the island towards the mainland. When it blows in the opposite direction, from the shore to the island, it is often accompanied by rain, and this would make wind dispersal less likely though violent squalls without rain can happen at any time during the monsoon season. Wind dispersal may be of more limited significance in the composition of the flora than the island's proximity to the shore suggests.

D. Succession and the role of fire

Succession is an ecological concept implying that plant communities will, through modification of their habitat, change over a period of time until a steady state or climax is reached. The rate of change, the structure of the intermediate communities and the type of climax depend on a host of factors including topography, soil, climate, availability of propagules and fire.

The role of fire in the ecology of the savanna and monsoon forests of south-eastern New Guinea has been debated at length but without absolute agreement on its effect. Van Royen (1963), for instance, suggested that much of the savanna and grassland is maintained by natural fires while Heyligers (1965, 1966) considered that savanna and grassland are largely caused and maintained by man-made fires. When Eden (1974) considered the distribution of forest, savanna and grassland in south-eastern Papua New Guinea in relation to topography, rainfall and soils, he concluded that 'the occurrence of forest on some most limiting sites suggests that under natural conditions a forest climax would be capable of occupying most, if not all, parts of the area'. However, Eden did not believe that all savanna and grasslands were the result of human activity, and concluded that man's role has been to extend rather than initiate the development of savanna. He suggested that the present distribution of these types of vegetation has been influenced both by past fluctuations in climate (such as drier periods in the mid-Holocene) and more recent activities by man, including clearing forest for shifting cultivation and burning.

Fire is an annual event in the savannas of south-eastern New Guinea and by the end of the dry season almost all areas have been burnt at least once. In the dry season, the ground is liberally scattered with exfoliated *Eucalyptus* bark and dead grass, up to 1 m high, that regenerates from a rootstock each year. Both grass and bark are highly inflammable. After burning, the ground is bare, the tree stumps are blackened and the branches are scorched. The onset of rain produces new growth of *Eucalyptus* and grass and broad-leaved herbs but if a further dry spell follows, the vegetation may be burnt a second time. In the following wet season, the vegetation returns to more or less its state before the fire. Patches of monsoon forest within the savanna may be scorched at the edges but do not normally catch fire.

The overall effect of fire is to maintain the *Eucalyptus/Themeda* association as a permanent fire disclimax and to prevent the spread of fire-intolerant species from adjacent forest patches. Eden (1974) found that when areas of forest on the mainland were cleared for gardens and not burnt when abandoned, they developed a shrubby vegetation with *Macaranga tanarius*, *Phaleria* sp. and *Abroma angusta* which shaded out grasses. The likelihood of the deflection of succession by fire is then decreased. On the other hand, if a cleared area is burnt, grasses become permanently established, seedlings of woody plants that are not fire-tolerant are lost and a fire disclimax results.

Since its occupation by the University in 1969, Motupore Island has been largely protected from fire although there were severe fires in 1982 and 1992. The 1992 fire was exceptionally severe due to the 10-year build-up of inflammable material so the effects may have been somewhat different from those of annual burning elsewhere in the region. This fire burnt approximately 75% of the savanna leaving the forest patches and strand more or less untouched. In the two years following the fire these observations were made:

1. In the savanna, the above ground herbaceous vegetation was entirely removed and the ground was exposed.

2. Stems of *Eucalyptus* and other trees or shrubs were scorched or even destroyed entirely.

3. Forest patches were scorched at the edges but were not burnt.

4. Some *Eucalyptus* trees and most savanna shrubs (e.g. *Ficus opposita, Clerodendrum floribundum, Flueggea virosa*) were destroyed above ground level and regenerated from root-stocks; in the case of smaller shrubs, they reached flowering size during the following wet season (Plate 8).

5. Before the re-establishment of the dominant grasses, quick growing herbs (particularly legumes, but also labiates), previously present but uncommon, established an extensive ground cover. Some herbs appeared that had apparently not been present in the flora in the years immediately before the fire (e.g. *Cleome viscosa, Merremia tridentata, Evolvulus alsinoides, Diplocyclos palmatus, Zehneria* sp.) presumably germinating from the soil seed bank or seed rain from an external source.

6. By mid- to late wet season, these herbs had given way to recovering grass and *Acacia* seedlings had appeared.

7. At least one species (*Ipomea nil*) in the scrub and strand grew and flowered profusely after the fire but had not flowered, or only very spasmodically, in previous years. The cause of this has not been investigated.

8. Sixteen months after the fire the dominant grass (*Themeda* species) had completely recovered its normal height (1.5-2 m) and was flowering. *Eucalyptus* regrowth had reached a height of 2 m and *Acacia* seedlings 0.5 m.

Approximately 65% of Motupore Island is currently savanna and it might be expected that in the absence of fire the whole island would eventually be covered by monsoon forest.

Before the severe fire in October, 1992, Menzies & Sengo (1993) looked for evidence of succession on the margins of the scrub patches. In most cases, savanna-scrub margins were abrupt with few signs of scrub species growing in the adjacent savanna. However, in one transect, *Bombax ceiba* saplings had proliferated at the upper margin (Fig. 4) but were absent lower down suggesting that perhaps the seedlings might not grow in shade. *Bombax ceiba* could then be regarded as one of the pioneer trees in the progression from savanna to forest. *Alstonia spectabilis* is also common in the northeast corner on the scrub border and the role of this species as a pioneer has already been mentioned.

Despite periods of protection from fire, succession on Motupore is likely to be a very slow process because of the severe seasonal climate and steep topography. Under present conditions, it seems likely that a light forest with a canopy of *Albizia* and *Garuga* with emergent *Bombax* is all that is likely to develop. Any increase in the diversity of the monsoon forest will be dependent on the availability of propagules from the mainland.

5. KEYS TO THE SPECIES

This key includes most of the trees, shrubs, climbers and broad-leaved herbs. However, ferns, grasses, grass-like herbs and cultivated plants and weeds growing in the station area are not included nor are plants not seen since 1985. A key to grasses will be found in Henty (1969).

The key is intended to be used in conjunction with the species accounts in the next section. Some key steps lead to a family or group rather than a single species because it is often simpler to compare the specimen in hand with the illustrations than to follow key steps through to a single species identification.

Abbreviations: fl = flower(s); fr = fruit(s); infl = inflorescence; lflt = leaflet(s); lvs = leaves.

KEY A - GENERAL KEY

1.	Mangrove tree with pencil-like aerial roots growing up from the mud	*Avicennia eucalyptifolia*
	Not like this	2
2.	Mangrove trees with pendulous, viviparous fr	Rhizophoraceae
	Not like this	3
3.	Mangrove tree with large white fl and flattened, spherical fr	*Sonneratia alba*
	Not like this	4
4.	Tree with single main stem, stilt roots and branches with terminal clusters of linear lvs	*Pandanus tectorius*
	Not like this	5
5.	Tree, usually with single main stem, prominent spines on trunk, compound digitate lvs and large red fl	*Bombax ceiba*
	Not like this	6
6.	Tree, usually with single main stem; lateral branches in whorls	7
	Not like this	8
7.	Latex absent; lvs clustered at branch ends; fr a large single-seeded nut	*Terminalia catappa*
	Latex present; lvs in whorls of 4; fr a long pendent follicle	*Alstonia spectabilis*
8.	Woody plant with a single, thick, unbranched stem to 2 m; crown of hard, pinnate lvs	*Cycas campestris*
	Not like this	9
9.	Leafless parasite twining round other plants	*Cassytha filiformis*
	Not a leafless twining parasite	10

10.	Leafless, terrestrial orchid with spikes of pink to purple fls	*Dipodium punctatum*
	Not a leafless, terrestrial orchid	11
11.	Vines or shrubs with tendrils	KEY B
	Not vines with tendrils	12
12.	Plants with opposite lvs	KEY C
	Plants with alternate lvs	13
13.	Plants with alternate compound lvs	KEY D
	Plants with alternate simple lvs	KEY E

KEY B - VINES OR SHRUBS WITH TENDRILS

1.	Shrub with 2-partite lvs	*Bauhinia binata*
	Vines	2
2.	Lvs linear, prolonged into tendrils	*Flagellaria indica*
	Lvs not linear and not prolonged into tendrils	3
3.	Annual herbs with soft stems	4
	Woody perennials	5
4.	Foetid-smelling herb with large white and purple bisexual fl	*Passiflora foetida*
	Not foetid-smelling; fl unisexual, small, white or yellow	Cucurbitaceae
5.	Spiny-stemmed vine with two tendrils at the node; veins parallel	*Smilax* sp
	Smooth stemmed vines with one tendril at the node; veins not parallel	Vitaceae

KEY C - LEAVES OPPOSITE ONE ANOTHER, EITHER SIMPLE OR COMPOUND.

1.	Lvs compound	2
	Lvs simple	4
2.	Vine with 3-foliate lvs and white fl	*Jasminum didymum*
	Not vines with these features	3
3.	Shrub with 3- to 5-foliate lvs; fl blue, in racemes	*Vitex trifolia*
	Herbs with 3-foliate lvs; fl white-yellow, in capitula	*Bidens* sp
4.	Vines	5
	Not vines	7
5.	Latex present	*Ichnocarpus frutescens;* Asclepiadaceae
	No latex	6
6.	Leaves with stipules	*Rhyssopteris timoriensis*
	Leaves without stipules	*Jasminum aemulum; Parsonsia quinquebullata*
7.	Herbs	8
	Shrubs or trees	15
8.	Stems square in section	9
	Stems terete	10

9. Leaf margins dentate; fl conspicuous, white, blue or purple — Labiatae
Leaf margins entire; fl minute and concealed below scarious bracts — Amaranthaceae

10. Succulent, maritime herb — *Sesuvium portulacastrum*
Not a succulent, maritime herb — 11

11. Lvs in unequal pairs, one large, one small; fl small, bluish-pinkish — *Boerhavia* spp
Lvs in equal pairs; fl all colours — 12

12. Fl solitary, showy, blue — *Dipteracanthus bracteatus*
Fl in infl, white, yellow or blue — 13

13. Fl microscopic, greenish, petals absent; whitish bracts sometimes present; sap often milky — herbaceous *Euphorbia* spp
Fl small, petals present; sap never milky — 14

14. Infl a capitulum; fl white or yellow — *Eclipta prostrata; Synedrella nodiflora; Tridax procumbens*
Infl a terminal raceme; fl blue — *Stachytarpheta jamaicensis*

15. Spiny shrub or woody herb with yellow fl — *Barleria lupulina*
Not spiny — 16

16. Shrub with very rough lvs; fr a fig — *Ficus opposita*
Lvs not rough; fr not a fig — 17

17. Small tree with yellow fl; fr a small, black berry — *Chionanthus ramiflorus*
Trees or shrubs with white fl; fr various — 18

18. Lvs orbicular, shiny, coriaceous; fl numerous, in a flat-topped infl — *Canthium suborbiculare*
Lvs and fls not as above — 19

19. Bushy tree; lvs narrow, obovate; Eucalypt-scented; behind mangroves on E side — *Osbornia octodonta*
Not like this — 20

20. Small shrub; lvs thick; dull green; fr fleshy, purple — *Jossinia desmantha*
Shrubs or trees; lvs not thick — 21

21. Shrubs; young stems and petioles red-purple; stamens long exserted — *Clerodendrum* spp
Shrubs or trees; young stems and petioles not red-purple; stamens ± enclosed in the fl — 22

22. Lvs large (>10 cm), ovate-orbicular; fl large (> 2cm); fr spherical, green; confined to strand — *Guettarda speciosa*
Lvs smaller, never orbicular — 23

23. Infl and fr compound — *Morinda citrifolia*
Infl and fr not compound — 24

24. Shrubs, infl terminal — 25
Trees or shrubs, infl axillary — 26

25. Flowering stems whitish; fl actinomorphic; fr red — *Psychotria bracteosa*
Flowering stems not whitish; fl zygomorphic; fr small, black — *Premna* spp

26. Dioecious tree or shrub; fr green to pale brown — *Timonius timon*
Monoecious tree; fr red — *Aidia cochinchinensis*

KEY D - LEAVES ALTERNATE, COMPOUND

1. Lvs 2-pinnate	2
Lvs 1-pinnate or 2- or 3-foliate	5
2. Scrambling shrub with paired spines at nodes	*Caesalpinia bonduc*
Erect shrubs or trees	3
3. Young stems and petioles rusty-pubescent; fl zygomorphic, yellow; fr usually one-seeded	*Peltophorum pterocarpum*
Young stems not rusty; fl actinomorphic, white; fr multi-seeded	4
4. Tree; infl an umbel of <20 fl	*Albizia carrii*
Shrub; infl ball-shaped with numerous fl	*Leucaena leucocephala*
5. Vines or twining herbs	6
Herbs, shrubs or trees	7
6. Lflt 5-7 pairs; fl large (>20 mm wide), solitary, pale blue	*Clitoria ternatea*
Lflt more than 10 pairs; fl small (<20 mm), dark blue-purple, in axillary racemes	*Abrus precatorius*
7. Herbs	8
Shrubs or trees	19
8. Lvs 3- or 5-palmate; fl with 4 yellow petals	*Cleome viscosa*
Lvs pinnate or 3-foliate; petals 4 or 5, various colours	9
9. Lvs pinnate	10
Lvs 3-foliate	13
10. Lflt very numerous (30-40 pairs), about 5 mm long; fl yellow	*Chamaecrista nictitans*
Lflt less than 15 pairs or, if more numerous, fl mottled red, brown and yellow and fr >10 cm long	11
11. Woody herb to 1 m high, all over whitish pubescent; fl white-pale yellow	*Tephrosia astragaloides*
Generally not woody and less than 1 m high; fl not white	12
12. Lflt up to 30 pairs; fl mottled red-brown-yellow; fr up to 15 cm long	*Sesbania cannabina*
Lflt generally fewer than 10 pairs; fl unicoloured; fr much smaller	*Chamaecrista absus; Indigofera colutea, I. hirsuta, I. linnaei; Tephrosia filipes*
13. Upright herb; fls in dense terminal infl	*Desmodium velutinum*
Twining or prostrate herbs; infl lax, fls separated	14
14. Robust perennial herb with woody pods to 10 cm long	*Canavalia papuana*
More delicate, annual herbs with much smaller pods	15
15. Fl dark red, almost black	*Macroptilium atropurpureum*
Fl never that colour	16
16. Lflt broader than long	*Christia obcordata*; *Rynchosia minima*
Lflt ovate, longer than broad	17
17. Fl orange and green	*Atylosia scarabaeoides*
Fl pink to blue	18

18.	Stems covered in long ferruginous hairs	*Glycine tomentella*
	Stems sparsely hairy or glabrous	*Galactia tenuiflora; Vigna radiata*
19.	Large deciduous tree; lvs pinnate, velutinous, the proximal leaflets with small, leafy «ears»; fr blue-black, 3-4 lobed, in hanging bunches	*Garuga floribunda*
	Shrub or tree without the above combination of characters	20
20.	Leguminous shrubs and trees; fr an elongate pod	21
	Not legumes; fr not elongate pods	25
21.	Lvs pinnate	22
	Lvs 2- or 3-foliate	23
22.	Lvs with >15 lflts; fls pale yellow; fr moniliform, many-seeded	*Sophora tomentosa*
	Lvs with 5-7 lflts; fls magenta; fr flat, few-seeded	*Pongamia pinnata*
23.	Lvs 2-foliate	*Bauhinia binata*
	Lvs 3-foliate	24
24.	Small tree; fl white; terminal lflt ovate; fruit breaking into 2-7 articles	*Desmodium umbellatum*
	Large tree; fl red; terminal lflt ± rhomboidal; fr cylindrical or moniliform, not breaking into articles	*Erythrina variegata*
25.	Lflt more than 5 pairs	26
	Lflt 5 pairs or less	27
26.	Lvs clustered at branch ends; fl minute; fr a large pendent orange capsule opening to reveal black seeds	*Cupaniopsis curvidens*
	Lvs not clustered; fl larger, white; fr upright, single seeded, succulent, red	*Micromelum minutum*
27.	Scrambling spiny shrub	*Harrisonia brownii*
	Upright shrubs or trees or, if scrambling, not spiny	28
28.	Lflt revolute, softly pubescent; fr green, tightly clustered, opening to reveal black seeds and crimson arils	*Alectryon repandodentatus*
	Lflt not revolute, glabrous; fr not as above	29
29.	Strand or mangrove tree with large (>10 mm) pendent white fl; fr large (>30 mm), green, spherical	*Xylocarpus* spp
	Not strand or mangrove trees; fl and fr smaller	30
30.	Shrub to 2 m high; lflt to 3 cm long; fr a small, red succulent berry	*Murraya paniculata*
	Small trees with larger lflt; ripe fr pink or orange	31
31.	Lflt 3-5	32
	Lflt 2-4 pairs	33
32.	Lvs with numerous oil glands; fr ripening orange	*Glycosmis cyanocarpa*
	Lvs without oil glands; fr ripening red	*Allophylus cobbe*
33.	Lflt apex acute; fr 3-lobed, ripening pink	*Ellatostachys tetraporandra*
	Lflt apex emarginate; fr bilobed, orange	*Harpullia leptococca*

KEY E - LEAVES ALTERNATE OR SPIRALLY ARRANGED, SIMPLE

1. Vines, lianes, trailing or twining plants — 2
   Not vines, etc. — 8

2. Climbing plants with thorny stems — 3
   Climbing plants without thorns — 5

3. Thorns on stems and petioles — *Dioscorea esculenta*
   Thorns only at the base of the petiole — 4

4. 2 down-curved thorns at the base of the petiole — *Capparis quiniflora; C. sepiaria*
   1 down-curved thorn at the base of the petiole — *Maclura cochinchinensis*

5. Twining climbers — 6
   Vines and lianes not twining — 7

6. Fl (rare) minute, green, unisexual — *Dioscorea bulbifera*
   Fl conspicuous, coloured or white, bisexual — *Ipomoea macrantha, I. nil; Jacquemontia paniculata; Merremia tridentata*

7. Trailing seashore plant with large purple fl — *Ipomoea pes-caprae*
   Woody forest liane; lvs rough; sap milky; fr red, succulent, several embedded in a warty base — *Trophis scandens*

8. Herbs, occasionally woody at the base — 9
   Shrubs and trees — 20

9. Infl a capitulum — *Emilia sonchifolia; Sonchus oleraceus; Vernonia cinerea*
   Infl various, not a capitulum — 10

10. Slender, upright annual; fl, white, solitary, axillary — *Evolvulus alsinoides*
    Not so — 11

11. Fl white, minute, in pedunculate axillary clusters — *Fatoua pilosa*
    Fl larger, not white — 12

12. Stems covered by sheathing leaf bases — 13
    Stems with leaf bases not sheathing — 14

13. Trailing herb with linear lvs; fls blue; common among grass — *Commelina ensifolia*
    Upright woody herb with ovate lvs; fl mottled yellow-brown; only on cliffs — *Dendrobium discolor*

14. Stems, thick, succulent; sap milky; lvs small, often striped white; fl red — *Pedilanthus tithymaloides*
    Not like this — 15

15. Legumes with zygomorphic, papilionoid fl — 16
    Not legumes; fl actinomorphic — 19

16. Delicate trailing or upright glabrous herb with orange or purple fl — *Alysicarpus* spp
    Upright herbs entirely covered with white hairs — 17

17. Lvs linear, only 2 mm wide; fl minute, red, in sparse axillary clusters — *Indigofera linifolia*
    Broad-leaved herbs; fl larger — 18

18. Lvs petiolate, ovate; fl in dense axillary clusters, purple — *Psoralea badocana*
    Lvs ± sessile, oblong-elliptic; fl in elongate terminal racemes, yellow — *Crotalaria montana*

19. Petioles long; stipules present, unequal; fl yellow — *Corchorus aestuans*
    Lvs sessile, clasping the stem; stipules absent; fl blue — *Plumbago zeylanica*

20.	Savanna tree with falcate, parallel-veined leaf-like phyllodes; fl yellow	*Acacia auriculiformis*
	Not so	21
21.	Tree (strangling when adult) with numerous prop roots; sap milky; apical buds covered by elongate sheaths	*Ficus microcarpa*
	Not so	22
22.	Mangrove tree with large (>20 cm) lvs, cordate at base and silvery below	*Heritiera littoralis*
	Not a mangrove tree with these characters	23
23.	Strand shrub with 2-partite lvs; fl white; fr a legume	*Bauhinia binata*
	Not so	24
24.	Dwarf shrub with thick stems and terminal clusters of lvs; sap milky; only on rocky headlands	*Euphorbia plumeroides*
	Not so	25
25.	Savanna tree with exfoliating bark and exuding gum	*Eucalyptus* spp
	Not so	26
26.	Tree with thick, stiff dull green lvs and parallel veins; bark thick, deeply furrowed; fr a yellow drupe on a succulent pink base	*Exocarpus latifolius*
	Not so	27
27.	Tree or shrub with distichous lvs, 2 prominent lateral veins running from base to apex of the leaf	*Celtis philippinensis*
	Not so	28
28.	Shrubs or small trees growing within upper tidal limits	29
	Shrubs or trees not growing in tidal zone	31
29.	Thick stemmed dwarf tree on rocky W shore; lvs with long, sheathing petioles	*Aegialitis annulata*
	Larger shrubs or trees without these leaf characters	30
30.	Lvs obovate or oblanceolate, emarginate; fr ovoid, <2 cm long	*Lumnitzera racemosa*
	Lvs elliptic, acute; fr curved, >2 cm long	*Aegiceras corniculatum*
31.	Trees or shrubs with minute, often unisexual, greenish fl, single or in small axillary clusters	32
	Trees or shrubs with conspicuous bisexual fl; infl various	38
32.	Male infl terminal, at first cone-like, later extending; female infl shorter; fr 3-lobed, green	*Excoecaria agallocha*
	Male infl not cone-like or terminal; fls uni- or bisexual	33
33.	Lvs thick and stiff; fl borne singly; fr succulent, red	*Drypetes australasica*
	Not so	34
34.	Tree; fl axillary, single; fr a black berry	*Bridelia oligantha*
	Trees or shrubs	35
35.	Shrub or small tree; lvs usually long-petiolate, 2 glands at lamina-petiole junction; fr a 3-partite capsule	*Mallotus philippensis*
	Shrubs; lvs with short petioles, without glands; fr fleshy	36
36.	Lvs distichous	37
	Lvs not distichous, grey-green; fr a white berry, several per axil	*Flueggea virosa*

37. Margin entire; fl one to several per axil, fr a pink berry — *Breynia cernua*
Margin serrulate; fl many in axillary infl; fr a minute orange drupe — *Trema cannabina*

38. Shrubs entirely covered in short, white hairs — 39
Trees or shrubs not covered in white hairs — 40

39. Lvs ovate, irregularly dentate; petiole short; fl white; fr a white berry — *Tournefortia muelleri*
Lvs cordate, minutely crenulate; petiole long; fl yellow; fr a capsule — *Abutilon indicum*

40. Shrubs or small trees with conspicuous, orange or yellow fl — 41
Shrubs or trees with white fl — 43

41. Lvs narrowly ovate; petiole short; fl yellow-orange, petals free — *Grewia* spp
Lvs much broader; petiole long; petals joined — 42

42. Lvs triangular to orbicular, markedly cordate; fl yellow with red throat; filaments fused to form a tube — *Hibiscus tiliaceous; Thespesia populnea*
Lvs not cordate; fl orange; filaments not fused — *Cordia subcordata*

43. Small shrub; lvs variable but often rhomboid and distichous, leathery; fl in short, axillary racemes; fr a red berry — *Scolopia novoguineensis*
Trees or shrubs without these characters — 44

44. Tree; lvs ovate, >10 cm long, faintly aromatic when crushed; fr a small, black berry — *Litsea glutinosa*
Trees or shrubs with smaller leaves — 45

45. Shrub or small tree; leaves distichous, soft and velvety; fl from short, leafless shoots; petals white-pale yellow, free; stamens fused; fr a capsule; seeds black; aril orange — *Turraea virens*
Trees or shrubs without soft, velvety leaves — 46

46. Lf margin crenulate or dentate — 47
Lf margin entire — 49

47. Tree with paired glands at junction of petiole and lamina, axillary racemes of sweet-smelling white fls; fr ovoid, blue when ripe — *Elaeocarpus arnhemicus*
Shrubs — 48

48. Lvs restricted to short lateral shoots; fl white; fr a triangular capsule ripening pinkish; seeds brown, arillate — *Maytenus emarginata*
Lvs not restricted to short lateral shoots; fl greenish, in axillary clusters; fr a spherical brown capsule; seeds black, no aril — *Colubrina asiatica*

49. Fls many, in much-branched terminal infl; fr 3 mm, red to black — *Ehretia cf dichotoma*
Fls solitary or few; fr larger — 50

50. Shrub or small multistemmed tree; petals 5, large, white; fr globose, brown, 3-4 cm, on a very long stalk — *Capparis lucida*
Single-stemmed tree; petals small, numerous, white, tipped brown; fr a red berry — *Mimusops elengi*

6. SPECIES ACCOUNTS

In the species accounts that follow, the vascular plants recorded from Motupore Island are arranged according to the major groups to which they belong, namely Pteridophytes (ferns), Gymnosperms (naked-seeds plants) and Angiosperms (flowering plants) which are further divided into Dicotyledons and Monocotyledons. The arrangement of the families within each major group is alphabetical, and genera and species are arranged alphabetically within their families. Some cultivated plants growing in the Station area are omitted.

For each species the Latin name is followed by the authority for that name. Synonyms are usually cited only when the alternative name has been used for specimens in UPNG herbarium or is in common use in the region. English names are given for a few species. References to key works where further information can be found are given in abbreviated form:

HFPNG - *Handbooks of the Flora of Papua New Guinea.* Vol. 1, J.S. Womersley (ed.) (1978); vol. 2, E.E. Henty (ed.) (1981). Melbourne University Press, Melbourne.

Fl. Mal. - *Flora Malesiana. Series I - Spermatophyta,* volumes 1-11. (1950-94). Flora Malesiana Foundation, Rijksherbarium, Leiden.

Henty - Henty, E.E. (1969). *A manual of the grasses of New Guinea.* Botany Bulletin No. 1. Department of Forests, Lae.

Henty & Pritchard - Henty, E.E. & Pritchard, G.H. (1982). *Weeds of New Guinea and their control.* Botany Bulletin No. 7. Department of Forests, Lae.

Frodin et al. - Frodin, D.G., Huxley, C.R. & Kirina, K.W. (1974). *Mangroves of the Port Moresby region.* University of Papua New Guinea, Biology Department Occasional Paper No. 3. (Revised version, 1982, by D.G. Frodin & G.J. Leach).

Millar - Millar, A. (1978). *Orchids of Papua New Guinea.* Australian National University, Canberra.

Verdcourt - Verdcourt, B. (1979). *A manual of New Guinea Legumes.* Botany Bulletin No. 11. Office of Forests, Lae.

For some taxa, there may be other relevant literature that is not cited here but this can be found in the reprint collection in UPNG herbarium or any botanical library.

Most plants are illustrated, the principal exceptions being mangroves, grasses and some weedy herbs, since illustrations and descriptions of these are available elsewhere (e.g. Frodin *et al.*, 1982; Henty, 1969; Henty & Pritchard, 1982). The drawings were done by the authors using fresh material collected on the island. This means that although some species are rather variable, the form present on Motupore is the one illustrated.

The descriptions only mention characters readily observed in the field and do not give full details of floral structure. Descriptions of the families and their key characteristics will be found in standard botanical works such as Mabberley (1987) and Heywood (1978). Characters obvious from the illustration (e.g. leaf size and shape) are not repeated in the text. Botanical terms are defined in the glossary.

The typical habitat on Motupore where the plant may be found is given, plus notes on status, where appropriate. For those plants that are not widespread, we indicate the location on the island where we found them in the years from 1990 to 1994. The wider geographical distribution of the species is summarised when known, but care must be taken in interpreting distributions when identity is uncertain or the taxonomy is confused.

For each species, the collections from Motupore in the UPNG herbarium are cited so that identifications may be checked and altered if incorrect or if the taxonomy changes. The identity of new material can be checked by comparison with herbarium specimens.

The following references may be useful for further information:

Brock, J. (1988). *Top End native plants.* Published by the author, Darwin.

Flora of Australia (1993). vol. 50. *Oceanic Islands 2.* Australian Government Publishing Service, Canberra.

Frodin, D.G. (No date). *Checklist of the plants of the Port Moresby District.* Card index, unpublished, UPNG herbarium; computerised version by Pius Piskaut and Lawong Balun.

Peekel, P.G. (1984). *Flora of the Bismarck Archipelago for naturalists.* Office of Forests, Lae.

Steenis, C.G.G.J.van & Veldkamp, J.F. (1987). Botanical guide to the sandy sea shores of Malesia. *Flora Malesiana Bulletin* 9 (4): 421-429.

Whistler, W.A. (1980). *Coastal flowers of the tropical Pacific.* Pacific Tropical Botanical Garden, Hawaii.

Notes:

The scale bar on all illustrations is equivalent to 1 cm unless otherwise indicated.

ASL = Above sea level.

HCFH collection numbers give number in HCFH series first, followed by number in the UPNG series. JM collection numbers give year of collection followed by number in JM series.

PTERIDOPHYTES

POLYPODIACEAE

Drynaria quercifolia (L.) J. Sm.

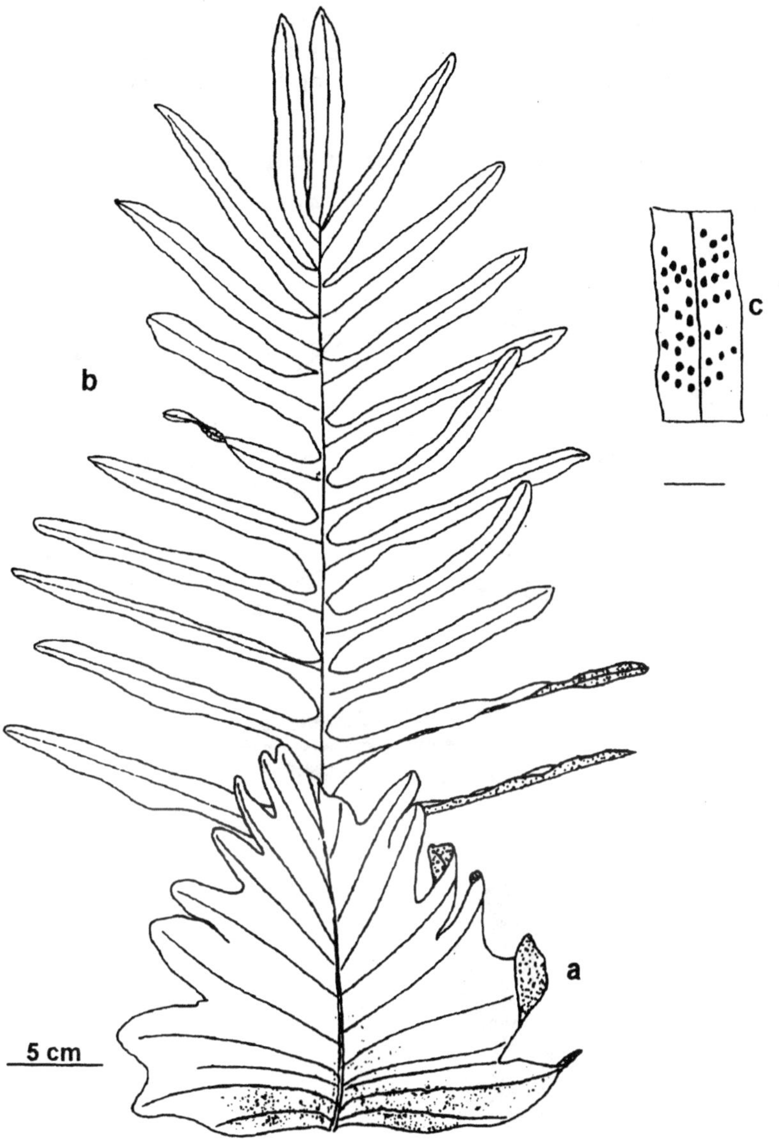

Fern with two types of frond: short, brown, basket fronds at base (a), which trap debris; longer, green, photosynthetic and sometimes fertile fronds (b) behind. Sori in rows, on lower side of green fronds only (c).

Abundant on ground, on rocks in scrub and as an epiphyte especially on cycads.

Sri Lanka to New Britain and N Australia.

HCFH 2026/13778; Maniwave TMM 39.

Pyrrosia longifolia (Burm.f.) Morton

Rhizome dark brown with paler scales Fronds leathery, strap-shaped (a), mid-green shiny above, covered in white scales below, except for dark mid-rib. Sori on lower side of photosynthetic fronds, along margins towards apex only (b).

Rare epiphyte on cycad in scrub, E side of island.

Malay Peninsula to Polynesia and N. Australia.

HCFH 2025/13777.

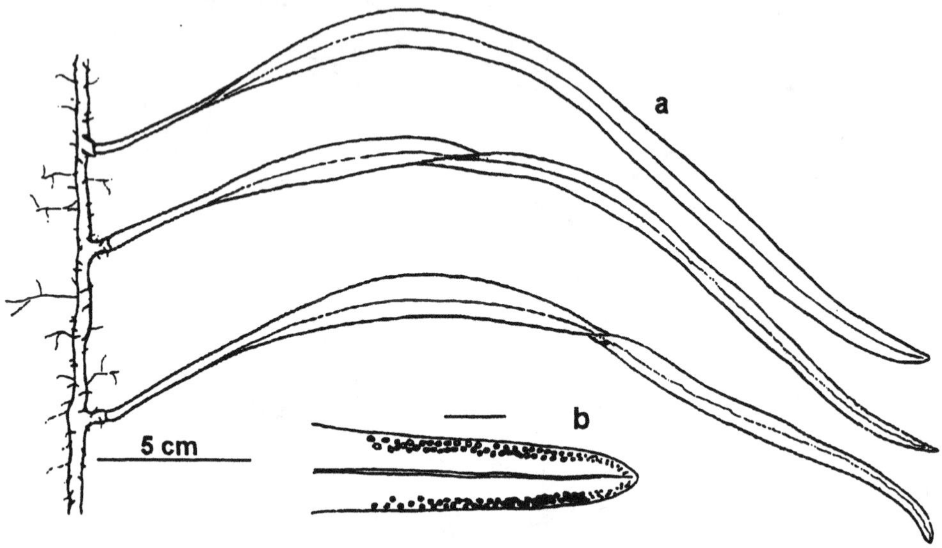

A third species of fern (c), not yet identified, grows in crevices on steep rock faces on the northwest side. Fronds bipinnate, to 10 cm, rachis wiry, black; leaflets blue-green above, rusty brown below; tomentose on all surfaces.

JM 95/147.

CYCADACEAE

Cycas campestris K.D.Hill

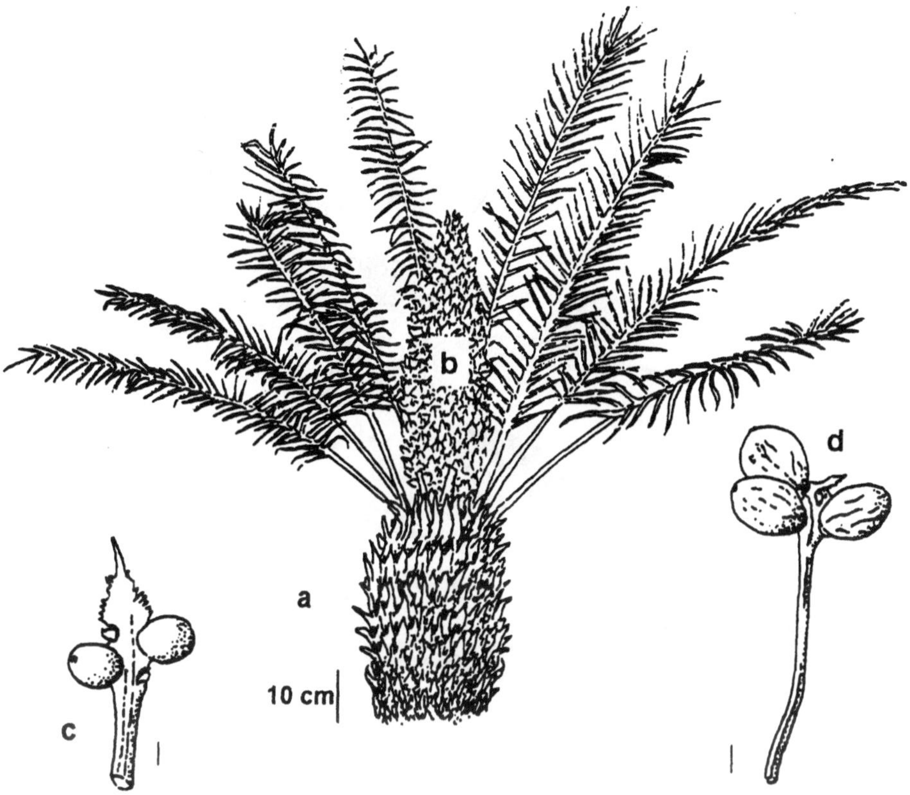

Palm-like tree with thick, usually unbranched, woody trunk, to 1.5 m high x 20 cm diameter. Leaves crowded into a rosette at apex of trunk (a), pinnate, to 2 m long, the leaflets dark green, hard; spines on petiole. Dioecious. Microsporophylls in brown elliptical cone to 30 cm long at apex of trunk (b). Megasporophylls (c) several, at apex of trunk, with 2-4 pairs of marginal ovules, green when immature, ripening red-brown, wrinkled (d).

In savanna and scrub. Frequent.

Endemic to the coastal plain around Port Moresby, often in areas prone to frequent fire.

HCFH 2031/13783.

ACANTHACEAE

Barleria lupulina Lindl.

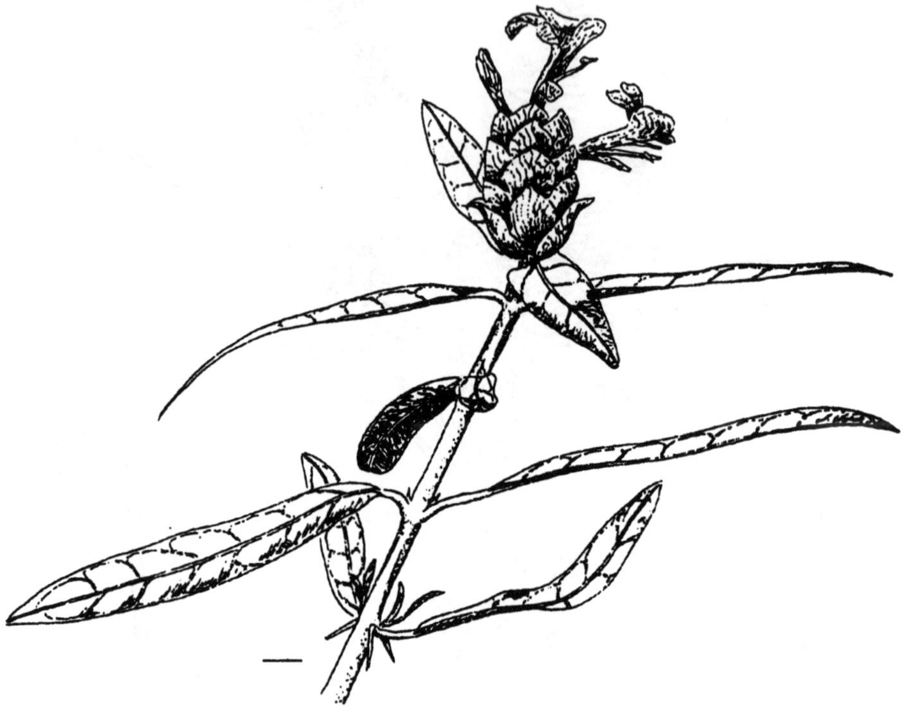

Rigid woody herb or shrub to 1-2 m tall. Leaves opposite, narrowly lanceolate, yellow-green above with a red mid-rib; thorns in 2 pairs at each node, down-turned. Inflorescence a terminal spike with large, papery, green to purple bracts. Flowers yellow, 5-lobed, tubular. Fruit a capsule.

Garden escape, invading grassland and forest edge near Caretaker's house and spreading up hillside.

Native to Mauritius; introduced to Papua New Guinea.

Frodin UPNG 3197.

Dipteracanthus bracteatus (R. Br.) Nees

Ruellia bracteata R.Br.

Herb, either prostrate or erect in more shaded places, to c. 20 cm tall. Leaves opposite, ovate, often with an acute apex, covered in fine white hairs. Flowers pale blue or mauve, in leaf axils, 5-lobed, tubular, borne in small numbers on each plant, ephemeral, opening early am and shed by mid am. Fruit an oval capsule.

Common in grassy area near buildings and at forest edge.

Australia and New Guinea.

Frodin & Pernetta UPNG 6674; HCFH 1072/13634.

Sesuvium portulacastrum (L.) L.

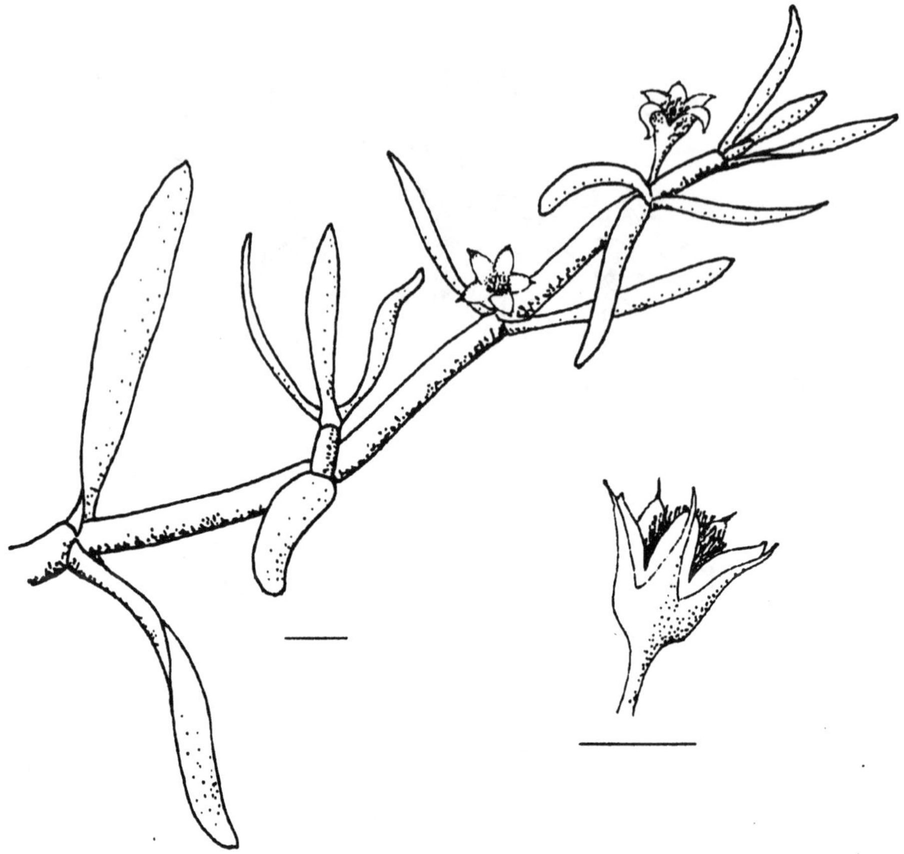

Succulent creeping herb. Leaves opposite, ribbon-like, fleshy, tasting of salt. Flowers purple, in leaf axils, 5-partite. Stems tinged red at nodes.

Sandy beach, NE corner, behind mangroves.

Sea shores throughout tropical regions.

Frodin UPNG 5244; Kembi s.n. 22/7/75; JM 94/131.

Trianthema portulacastrum L. [not illustrated]

(Henty & Pritchard: 53)

Prostrate or ascending herb. Leaves opposite, somewhat fleshy, dull. Flowers small, pink.

Station area.

Pantropical.

Frodin UPNG 5232; Lambinon 87/93.

AMARANTHACEAE

Achryanthes aspera L. [not illustrated]

(HFPNG **1**:13. Henty & Pritchard: 54)

Herb to 70 cm with square stems. Leaves opposite, soft and woolly. Flowers small, in a elongate terminal spike, greenish. Fruit 1-seeded surrounded by persistent spiny bracteoles.

Occasional, on ridge top and elsewhere, in shade.

Old World tropics and subtropics.

JM 94/108.

Alternanthera pungens Kunth. [not illustrated]

(HFPNG **1**: 18; Henty & Pritchard: 56)

Prostrate herb. Leaves opposite. Flowers in prickly axillary clusters, green-yellow when young; tepals yellow, inside scarious bracts.

Station area.

Originally native to S America and introduced into palaeotropics. Naturalised in New Guinea and Australia.

Frodin UPNG 5228, 5230.

Alternanthera sessilis (L.) DC. [not illustrated]

(HFPNG **1**: 20; Henty & Pritchard: 57)

Prostrate herb. Leaves opposite. Flowers in sessile axillary clusters, whitish.

Station area.

Pantropical but possibly palaeotropical in origin.

Frodin & Pernetta UPNG 6661.

Celosia argentea L. [not illustrated]

(HFPNG **1**:30; Henty & Pritchard: 60)

Annual herb to 1.5 m with stiff, woody stems. Leaves small, rather distant, reddish. Inflorescence terminal, crowded. Perianth scarious, white, turning pink.

Occasional on ridge top, where previously burnt.

Pantropical, perhaps of African origin.

JM 94/107.

Cyathula prostrata (L.) Bl.

(HFPNG **1**: 32; Henty & Pritchard: 61).

Ascending herb to 1 m or more tall. Leaves dull green, glabrous, margin entire. Petioles, main veins and stems tinged red. Inflorescences terminal, in upper leaf axils. Flowers in clusters; modified sterile ones bearing spiny hooks (a). Fertile flowers (b) with 4 green tepals, with hairy margins; stamens 4. Fruit with one black seed.

Edge of monsoonal scrub. E and W sides of island, in shade on slope, 2 m ASL.

Palaeotropical; probably introduced into tropical America.

Frodin UPNG 5217; HCFH 1749/13681; Powell et al. UPNG 2334.

ANNONACEAE

Annona squamosa L. [not illustrated]

Custard Apple or Sweetsop.

Small, spreading, semi-deciduous tree, to 6 m. Leaves alternate, lanceolate, ± distichous; margins entire, involute; veins prominent below. Flowers solitary, pendulous, with 2 whorls each of 3 yellow-green petals, surrounding numerous short stamens and a composite stigma on a central dome. Fruit green, a syncarp; surface lobed; flesh white, sweet, containing numerous black seeds c. 1 cm long.

Two or three planted in small patch of trees near houses.

Native to tropical America.

HCFH 1766/13698.

APOCYNACEAE

Alstonia spectabilis R.Br.

Pagoda-shaped tree to 10-15 m high, with 4 ± horizontal branches at each node. Stems square. Contains white sap. Leaves (a) in whorls of 4, obovate, dark green above, yellow-green below, glossy; veins white, prominent. Inflorescence (b) cymose, pedunculate, ± hemispherical. Flowers (c) small; corolla tubular, 5-lobed, with hairs on exposed surface of lobes, white, furry. Fruit (d) a long, thin, pendent follicle, splitting longitudinally, green, turning brown at maturity, containing numerous brown seeds (e) each with tufts of white hair.

Monsoonal scrub and savanna woodland on hillside.

Malesia E to NG and Solomon Islands; also Queensland.

HCFH 932/13463, 1712/13644; Lambinon 87/135.

Alstonia spectabilis R.Br.

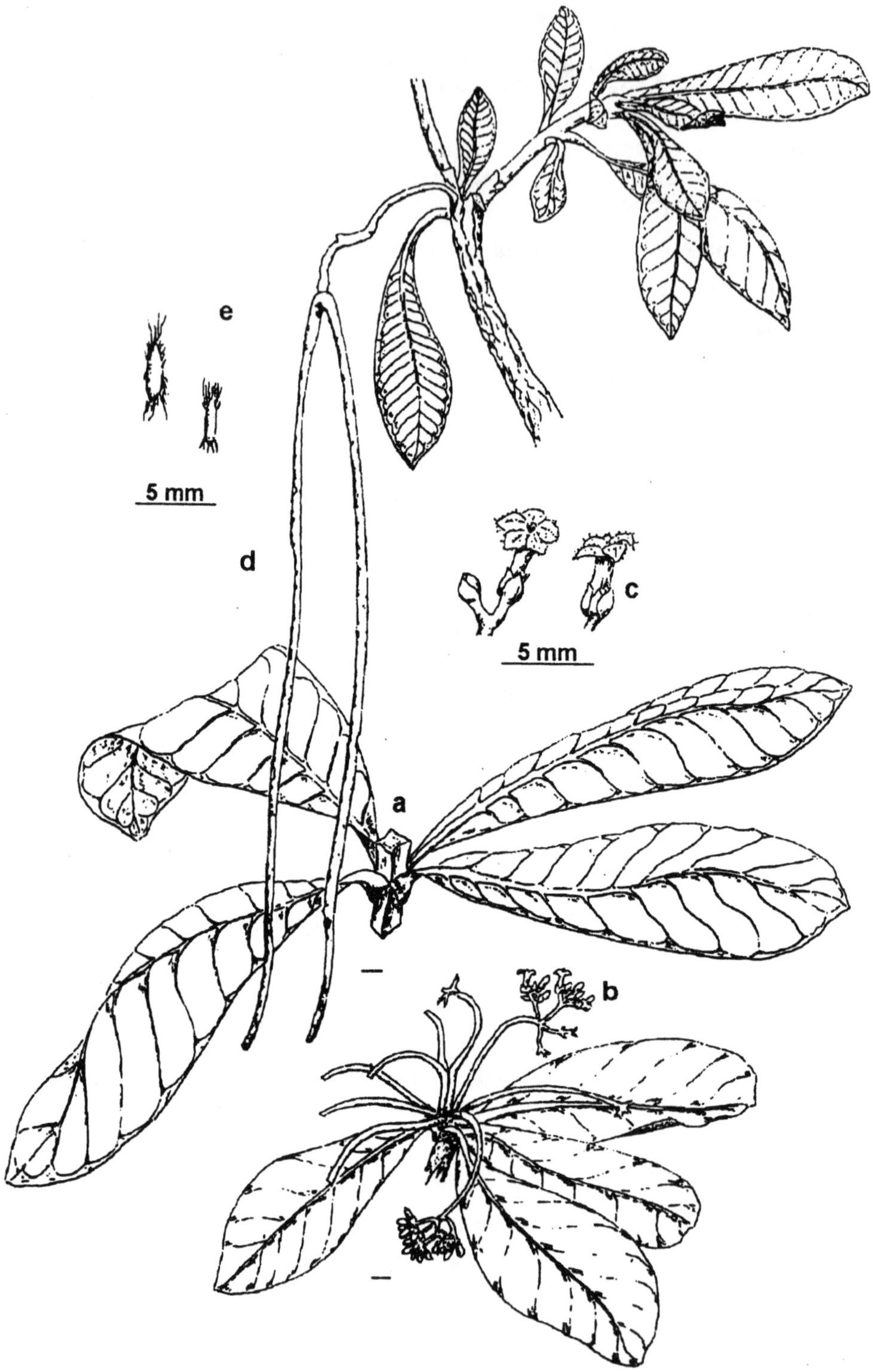

Catharanthus roseus (L.) G.Don [not illustrated]

(Henty: 16).

Herb, to 30 cm high. Leaves opposite, ovate. Flowers in upper leaf axils; corolla tubular, 5-lobed, pink. Fruits green with longitudiual grooves.

Planted near Old Hostel and escaped into open grassy area.

Native of Madagascar, cultivated widely in tropics.

HCFH 1700/13632.

Ichnocarpus frutescens (L.) W.T. Aiton.

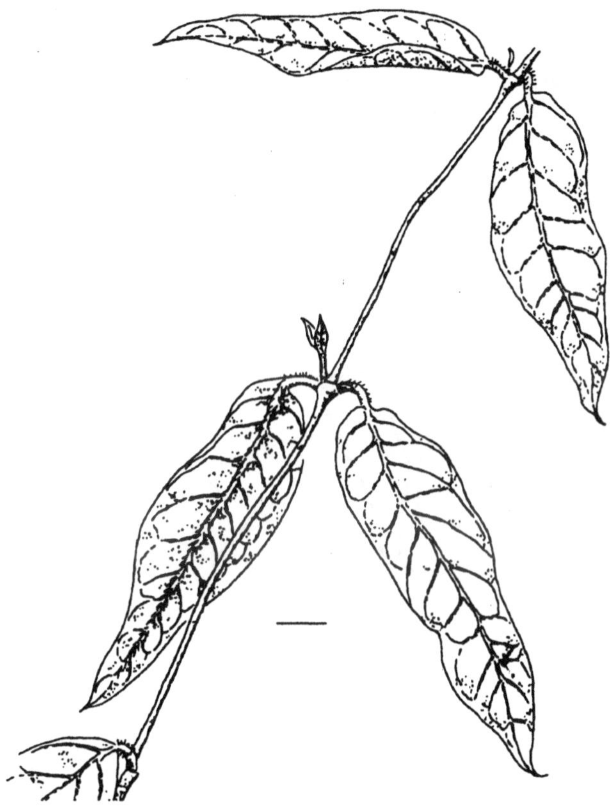

Vine. Stems brown-green, hispid due to minute spine-tipped warts; leaves thin, green with yellow hairs on midrib and petiole; latex present.

Forest edge, east slope. Common but rarely flowering.

Asia, Malesia and Australia.

HCFH 1031A/13563A; JM 92/21.

Parsonsia quinquebullata Sleesen

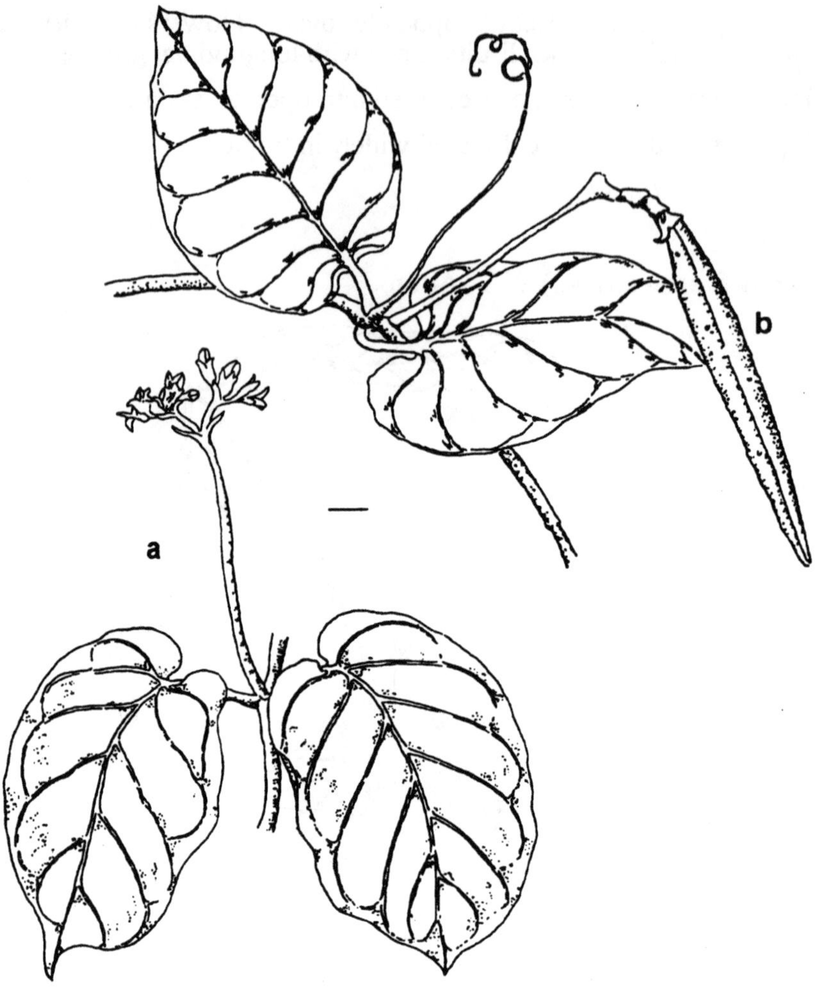

Creeper. Stems woody, brown with numerous small lenticels. Sap not milky. Leaves opposite, mid- to dark green and glossy above, velvety when young, sometimes less so with age. Inflorescence (a) axillary, pedunculate, cymose. Flowers dirty greenish yellow, tubular. Fruit (b) a long thin follicle, deeply grooved, grey-green, dehiscent.

Common in savanna.

Moluccas and New Guinea.

Frodin UPNG 827, 2931, 3182, 4275 & s.n.; HCFH 656/13187, 1029/13561, 1756/13688; Huxley s.n.

ASCLEPIADACEAE

Cynanchum carnosum (R.Br.) Schltr.

Ischnostemma carnosa (R.Br.) Merr. & Rolfe

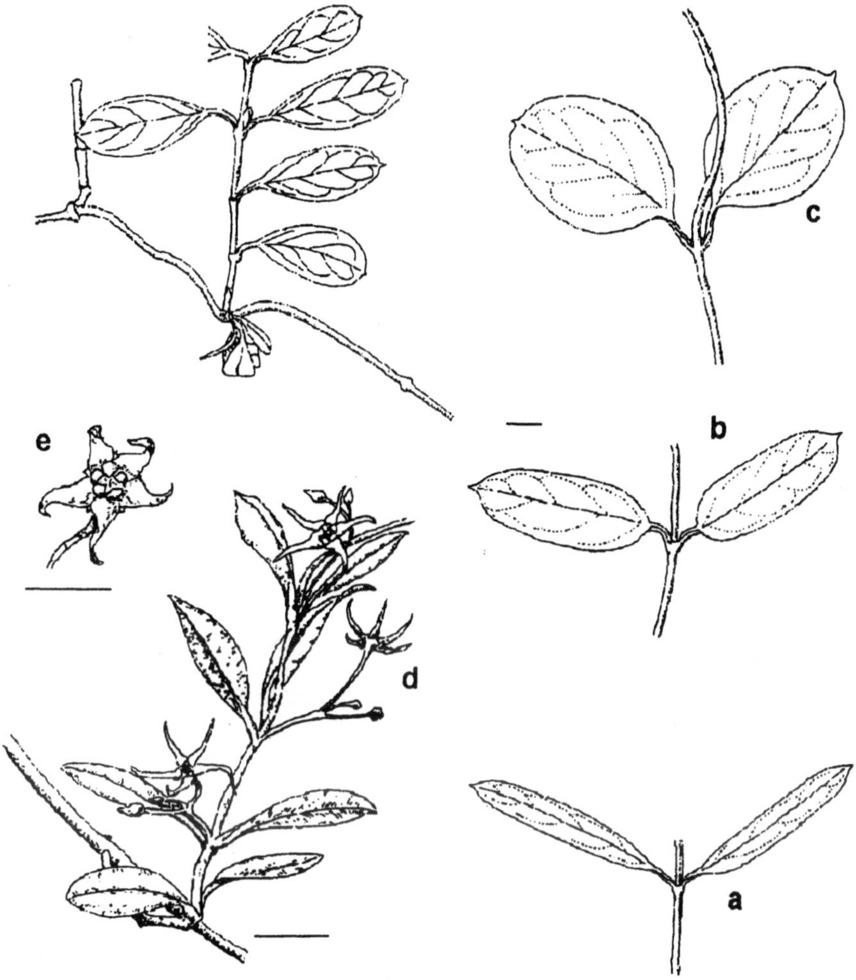

Creeper. Stems green. Sap milky. Leaves light green, slightly fleshy when fresh, variable in shape according to position on plant but nearly always mucronate: narrowly lanceolate at base (a), becoming elliptical (b) and ovate (c) towards top of plant. Flowers in axillary umbels (d), 5-partite, star-shaped (e), greenish yellow, inner segments white. Fruit a follicle containing numerous white silky hairs and brown seeds, each with a thin, membraneous margin.

Back of mangroves, E side of island, common.

Malesia, to N Australia.

Frodin UPNG 5240; HCFH 1039/13571, 1754/13686; JM 91/1; Vinas UPNG 4834.

Dischidia ovata Benth.

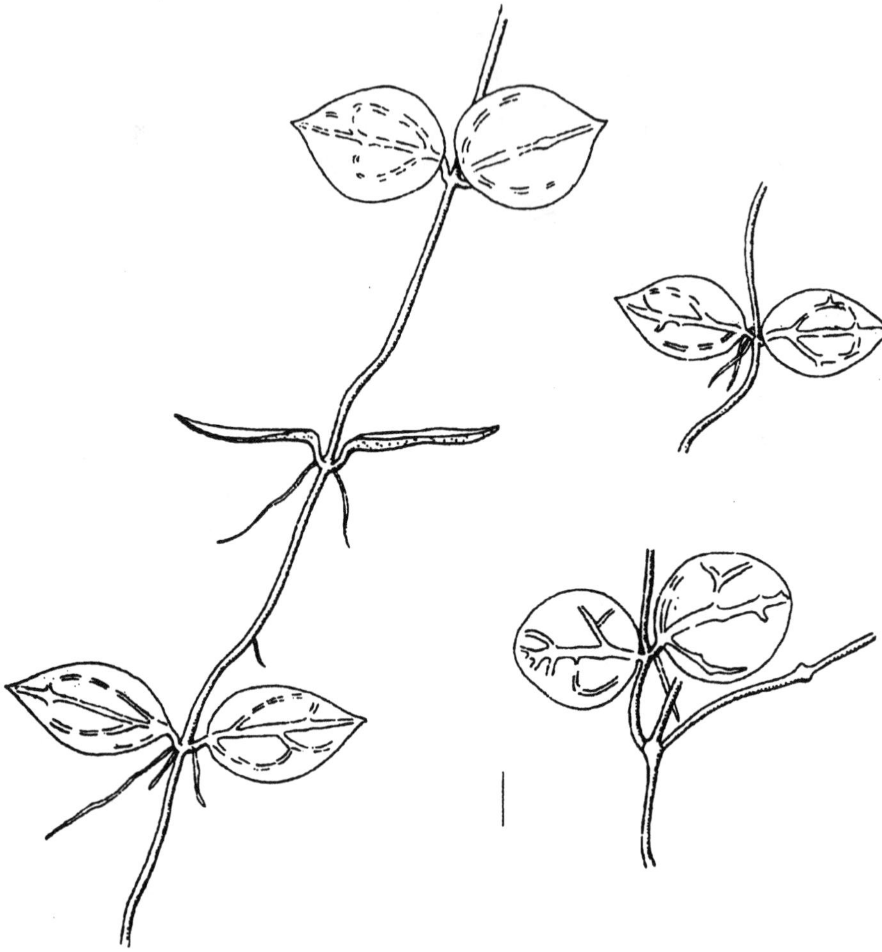

Lithophytic or usually epiphytic succulent herb, with creeping stems bearing adventitious roots at nodes. Latex white. Leaves thick-fleshy, stiff, reddish or green with white pattern of venation visible on upper surface only, scurfy. Inflorescences 1-8 flowered, axillary. Flowers small, 5-partite, stained reddish but not seen at Motupore. Fruit a follicle.

On rocks 3 m ASL on W side of island in sheltered position.

New Guinea and Queensland.

Bigilale 50; HCFH 652/13156, 1753/13685.

Gymnanthera oblonga (Burm. f.) P.S.Green

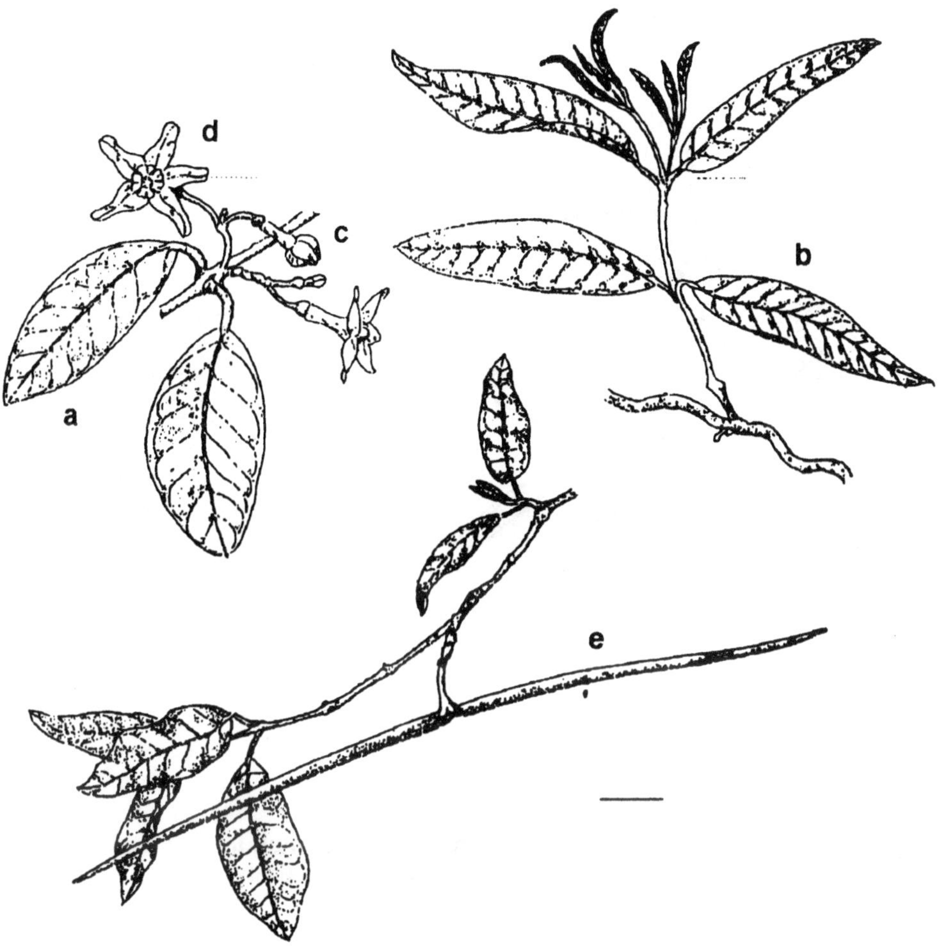

Creeper. Stems brown-green with prominent lenticels. Latex white. Mature leaves (a) elliptical, mid-green, very shiny above, dull below; venation pale when leaf held up to light. Petioles and younger leaves (b) reddish, especially venation. Flower buds elongate, expanded towards apex (c). Flowers in small axillary clusters, white (d). Fruit (e) a follicle with 2 elongate, pencil-shaped lobes, attached at base.

Monsoonal thicket, E and W sides of island, just above strand line, not common.

Malesia and Australia.

Frodin UPNG 4266; HCFH 1755/13687; JM 91/2, 93/60.

Gymnema geminatum R.Br.

Creeper. Stems brown-green with numerous small lenticels, shortly hairy. Latex white. Leaves dull green above, glabrous, shortly hairy below, not fleshy. Flowers creamy-white (a). Fruit a follicle, green.

Uncommon, at scrub margin. 1 m ASL.

New Guinea and Australia.

HCFH 1745/13677; JM 92/37.

Marsdenia velutina R.Br.

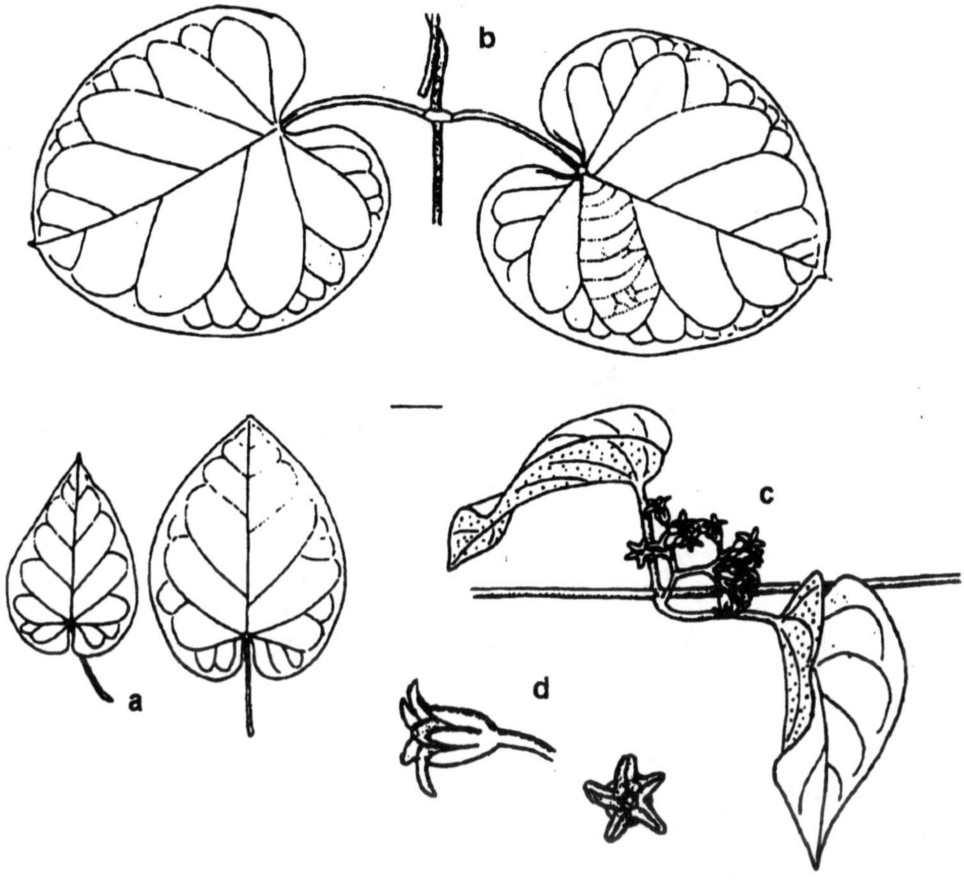

Woody creeper. Latex white. Leaves bright mid- to darkish green above, sometimes blotched, variable in size and shape from 7.5 x 5 cm to 12.5 x 12.5 cm, heart-shaped (a) to ± round with a cordate base (b), softly hairy to woolly on both sides; venation yellow, prominent below. Flowers in axillary clusters (c); corolla greenish, becoming orange-yellow, 5-lobed, fleshy (d) with a musky smell. Fruit a follicle, green, c. 4.5 x 2 cm.

Common and conspicuous in savanna, especially along ridge top.

Australia, Timor, Philippines and New Guinea.

Frodin et al. UPNG 2932; HCFH 654/13185, 1704/13636, 2024/13776.

Secamone elliptica R.Br.

Twining woody creeper with latex. Leaves yellow-green to darkish mid-green, glabrous above, variable in shape from narrowly lanceolate (a) through broadly lanceolate (b) to ovate (c), sometimes fleshy when old; venation not prominent. Flowers in axillary clusters (d), corolla 5-lobed, yellowish (e). Fruit a follicle, 2-lobed (f) but often only 1 lobe develops (g), green when immature turning brownish, ± angled. Seeds numerous, plumed (h).

Common in savanna and in canopy of monsoonal scrub, always in sunny positions.

India, E and SE Asia, Malesia, N Australia and New Caledonia.

Frodin et al. UPNG 2933; HCFH 655/13186, 1033/13565, 1705/13637, 1787/13719, 2010/13762; Huxley s.n.; Powell et al. UPNG 2327, 2339.

Bombax ceiba L.

Bombax malabaricum DC.

(HFPNG **2**: 5)

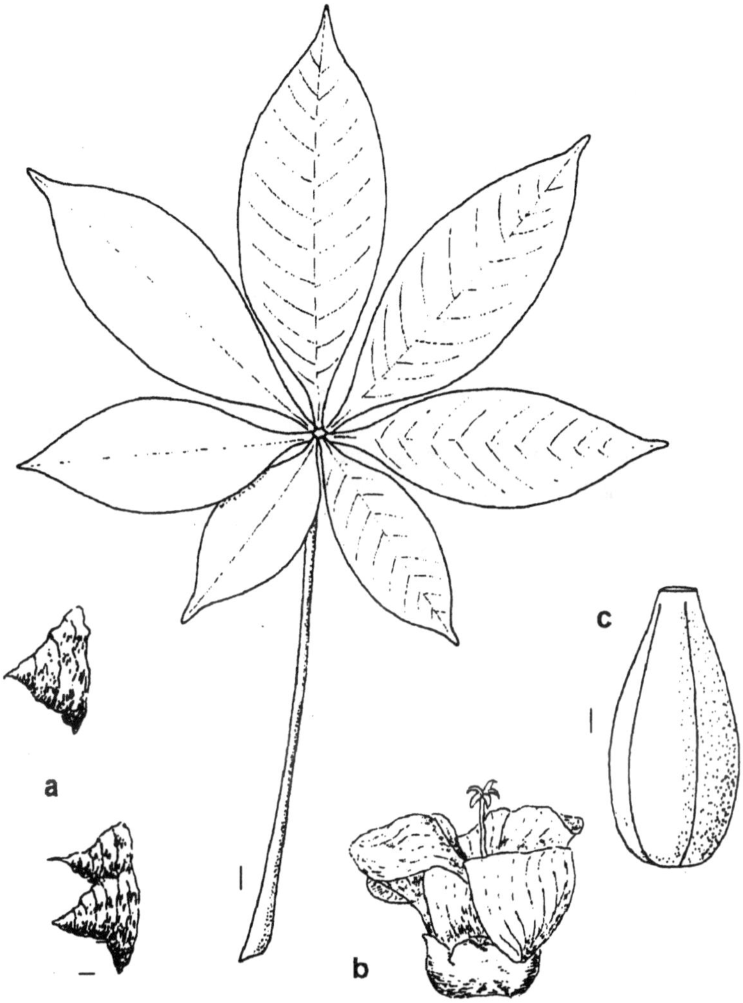

Deciduous tree, to 15+ m, flowering when leafless. Trunk covered with large, conical, corky, sharp-tipped spines (a). Leaves in clusters at ends of twigs, digitately compound with 5-7 leaflets. Flowers (b) cup-shaped; calyx green-yellow; petals 5, deep scarlet, thick-fleshy; stamens numerous, in 2-3 whorls, variable in length; style 1; stigmas 3. Fruit (c) a black capsule, pendulous, with 5 thin, woody valves, containing seeds embedded in cotton-like material.

Monsoonal woodland on lower hillside, east side only.

India and Ceylon through Malesia to Solomon Islands.

HCFH 1053/13585.

Cordia dichotoma Forster f.

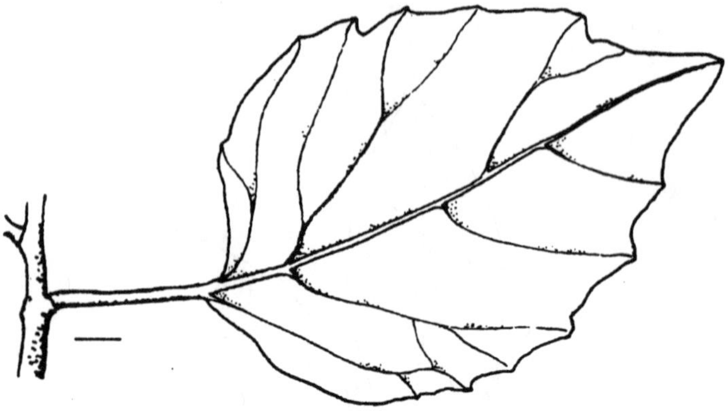

Small bushy tree. Leaves alternate; petioles to 3.5 cm; blade broadly ovate to ± circular, up to 10 x 8 cm, mid-green above, paler below, glabrous; upper surface drying with numerous tiny white dots (clusters of mineralised epidermal cells); margin crenate; secondary and tertiary venation reticulate. Flowers in branched, terminal clusters; corolla white. Fruit a drupe with smooth green exocarp and pink flesh, ± spherical, 1 cm diameter, with cup-like persistent calyx.

Occasional in scrub on E and W sides of island

India through S & E Asia, Malesia to NE Australia and New Caledonia.

Frodin UPNG 4260; Huxley s.n.; JM 95/143.

Ehretia cf dichotoma Blume [not illustrated]

Tree to 10 m with large spreading branches or scrambling shrub. Stems greenish to pale grey-brown. Leaves alternate; petiole to 1 cm long; blade mid-green or yellowish green, elliptic-lanceolate, glabrous; apex and base acute; margin entire. Inflorescence lax, terminal, much branched. Flowers not seen. Fruits small, spherical, 3 mm diameter, red to black.

Coastal scrub on steep slope, W side of island.

Ehretia dichotoma is known from Indo-China, Malay Peninsula, Andaman Islands and Java.

Frodin UPNG 2942; HCFH 1723/13655.

Cordia subcordata Lam.

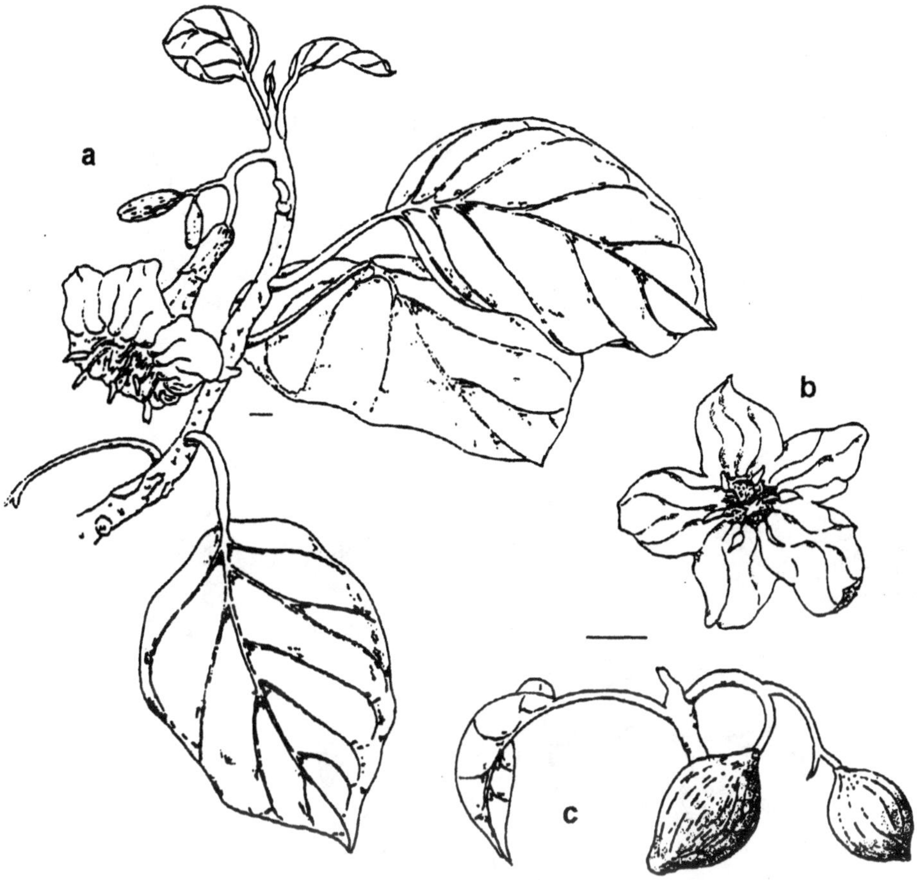

Small tree to 6 m. Young stems green, old ones brown, with a sudden transition. Petioles yellowish. Leaves mid-green, glossy above, paler below, glabrous; margin entire, slightly undulate. Flowers in a loose cyme (a); corolla tubular with 5-7 lobes, bright orange (b); stamens 6. Fruit a hard 'nut' (c); exocarp green, turning black and dry when ripe; mesocarp fibrous or corky; endocarp bony and ridged, with 4 cavities. A common drift fruit.

Common on strand line.

Coasts of E Africa, Indian Ocean, Indo-China, Malesia, S. Pacific Islands.

Frodin UPNG 5206; Powell et al. UPNG 2341..

Tournefortia muelleri I.M. Johnst.

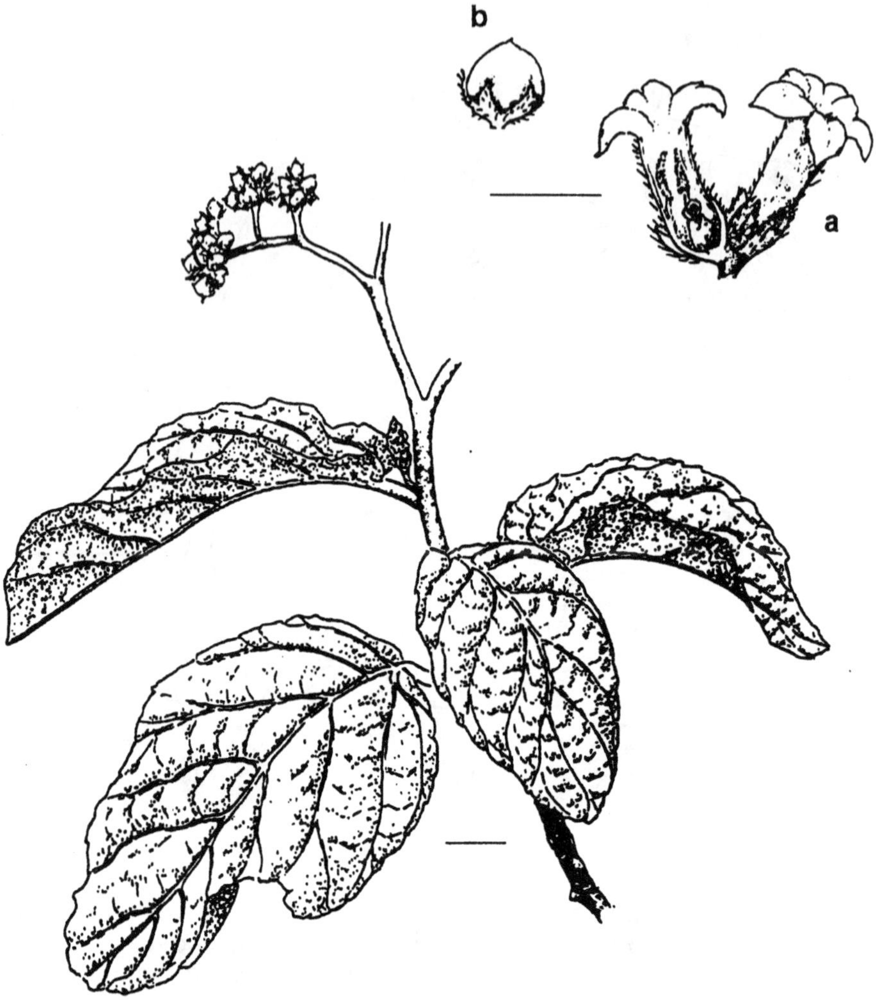

Shrub to c. 3 m high. Leaves alternate, shortly petiolate drying black; blade ovate, soft; margin irregularly dentate. Almost all vegetative parts and flowers densely covered in short white hairs. Flowers in dense heads, white (a). Fruit a soft white berry (b) with black seeds.

Rare, occasional plants on W side.

N Australia and S Papua New Guinea.

JM 91/14, 93/57.

Garuga floribunda Decne.

(Fl. Mal. 5: 215)

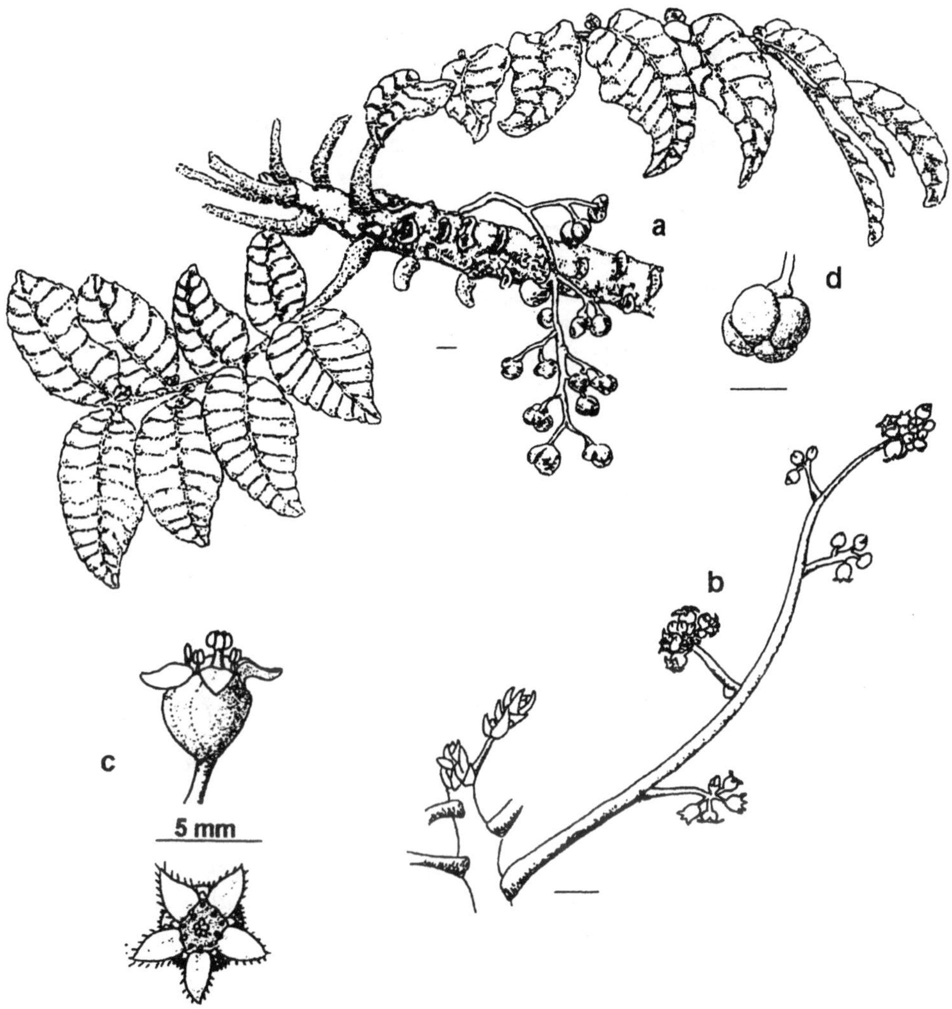

Large or medium tree with heavy spreading branches, deciduous from June to October, leaves turning red before falling. Leaves pinnate, clustered at ends of thickened brown twigs which bear conspicuous leaf scars (a); leaflets opposite, the lower ones with 'ears' at the base, pale yellow-green above, yellower below, velutinous on both sides and on rachis; margins serrate. Flowers on short pedicels in a lax panicle (b), small, yellow (c). Fruits (d) blue-green, ripening blue-black, irregularly (1-3) 4-lobed, with thick green mesocarp.

Common on both sides of island in monsoonal scrub.

Asia, Malesia to N Australia and W Pacific Islands.

Frodin UPNG 4258; Hehonah 4; HCFH 605/13136, 1741/13673; JM 94/135.

Capparis lucida (Banks ex DC.) Benth.

(Fl. Mal. **6**: 92)

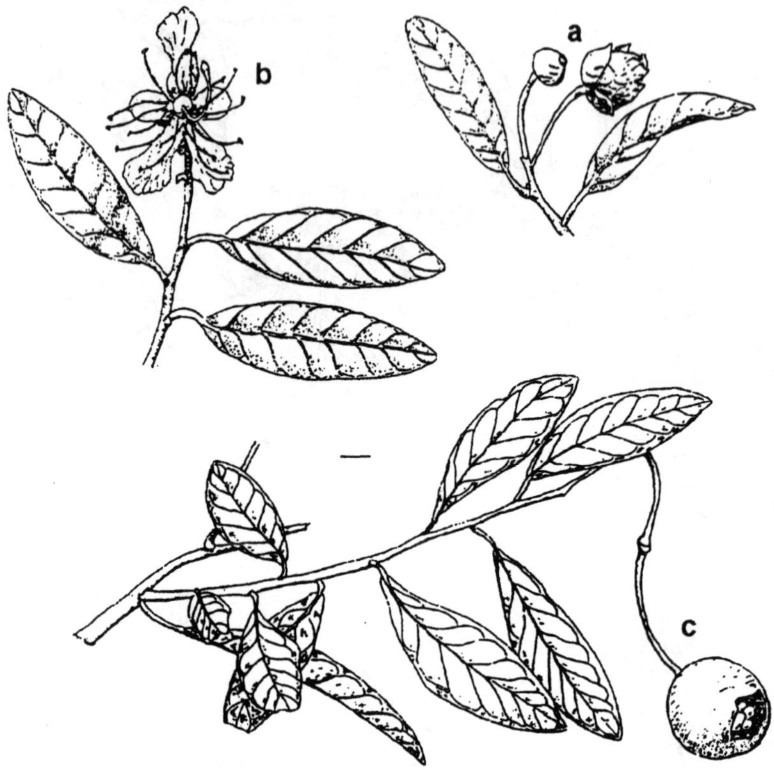

Large spreading shrub or small tree; deciduous. Leaves yellowish to mid-green above, mottled, dull or slightly glossy, paler and dull below, glabrous; venation not prominent. Young stems and petioles minutely hairy, green to brown, without thorns. Flower buas 1 or 2 per axil, 1.5 cm long (a). Flower showy (b); calyx lobes 2, green; corolla lobes 5, 2 smaller, white and green, 3 larger, creamy white, the lower one deeply lobed; stamens numerous; ovary far exserted. Fruit (c) green, becoming dark brown, 2-4.5 cm diameter, borne on a gynophore, itself borne on a pedicel; seeds numerous, small, kidney-shaped.

Coastal scrub on N and E side of island and on ridge top.

S Sulawesi, Lesser Sunda Islands, S New Guinea, NW Australia.

A. Elias s.n. 14/8/80; HCFH 653/13184, 1040/13572, 1722/13654, 1757/13689.

Capparis quiniflora DC.
(Fl. Mal. **6**: 89)

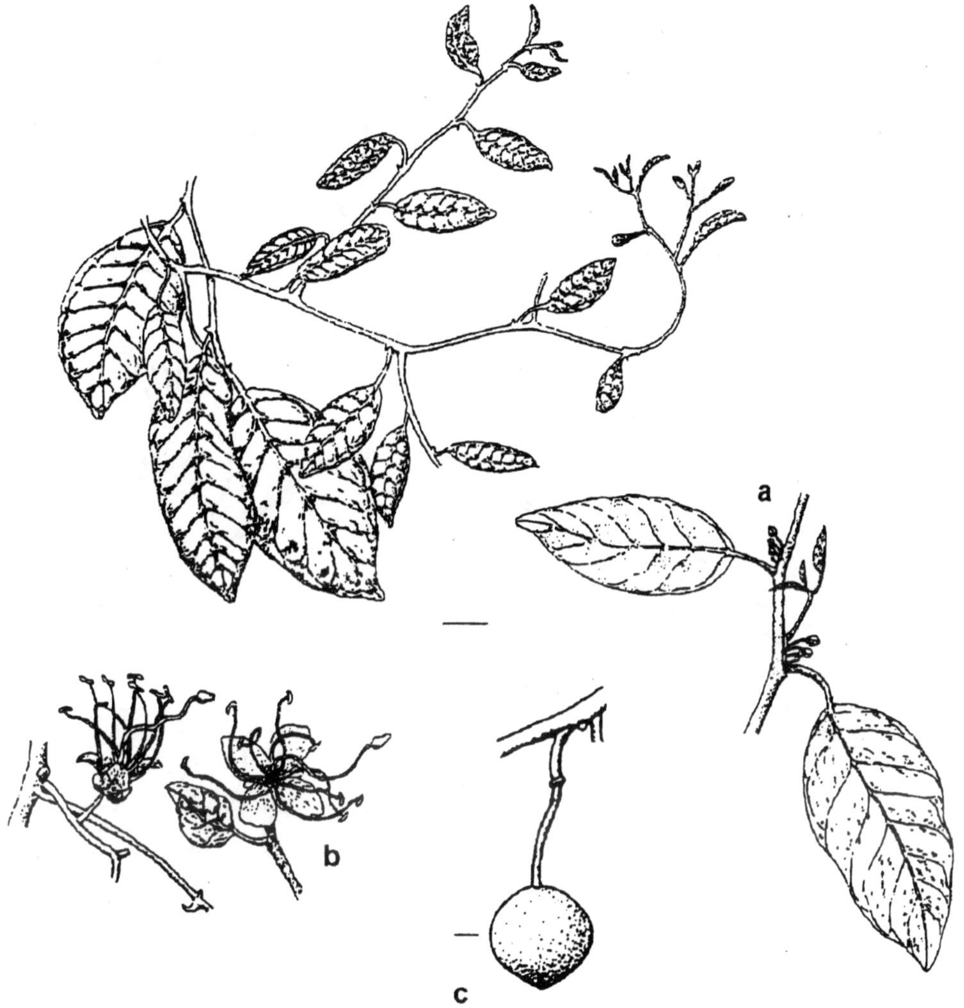

Climbing shrub. Stems green with raised area in each axil and frequently bearing two thorns at base of each petiole. Leaves dark green or yellow-green, semi-glossy above, hard, thick, with corrugated surface; young leaves olive-green, with white coating below, softer; venation faint. Flower buds c. 4 mm diameter, in a series in each axil (a). Flowers (b); sepals 4, green; petals 4, white; stamens 8, long; ovary far exserted. Fruit (c) red purple, c. 4-5 cm diameter, seeds numerous.

Coastal scrub, common on W slope, less common on E side.

S Sulawesi, Lesser Sunda Islands, Moluccas, NG, Fiji, New Caledonia, N Australia.

Frodin UPNG 3184, HCFH 1073/13605, 2006/13758; Huxley s.n.; JM 91/8; 94/129.

Capparis sepiaria L.
(Fl. Mal. **6**: 79)

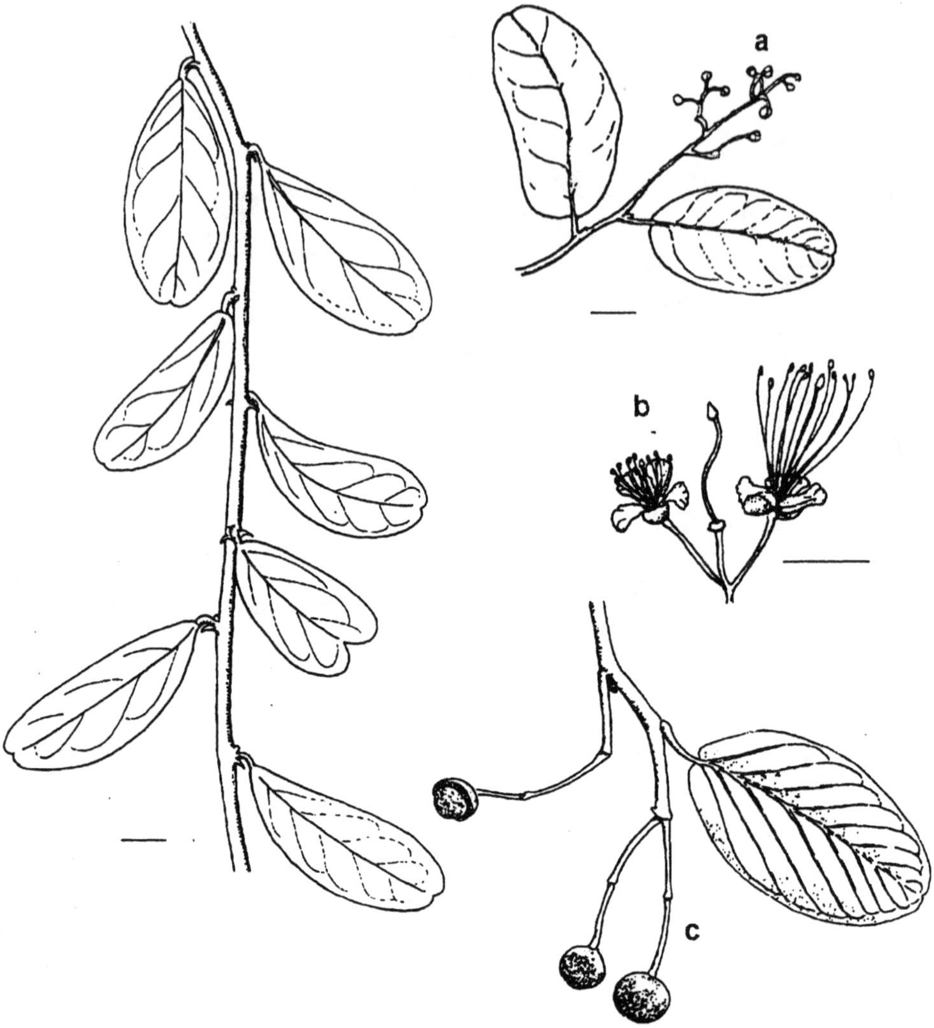

Scrambling woody vine or shrub. Leaves yellow-green or mid-green above, paler below, notched at apex. Two down-curved spines at base of some petioles. Buds c. 4 mm diameter. Inflorescence (a) terminal or axillary, branched. Flowers (b) similar to those of *C. quiniflora*. Fruit (c) green becoming purple-black, 1-1.5 cm diameter with a single seed.

Coastal scrub, E side of island.

India and Sri Lanka to New Guinea.

Frodin UPNG 815, 4489; HCFH 2003/13755;JM 94130.

Distinguished from *C. quiniflora* by the leaves being emarginate, rather than acute, at the apex, flowers on a branching inflorescence and stamens numerous, not 8.

Cleome viscosa L.

(Fl. Mal. **6**: 103; Henty & Pritchard: 64)

Quick-growing annual herb, to 2 m. Leaves 3- or 5-palmate. Flowers (a) solitary in axils; sepals 4, green; petals 4, free, yellow; stamens numerous; ovary single. Fruit (b and c) an elongate capsule opening by 2 valves, ridged when mature.

There are 2 varieties of this plant, one is taller, appearing glabrous, though covered with minute non-glandular hairs (b and d). The other is shorter, more robust and entirely covered in conspicuous, sticky glandular hairs (c).

Locally common in grassland on ridge and slopes, especially on burnt areas. Appeared after the fire in 1992.

Africa to Australia, naturalised in America.

JM 93/61.

CARYOPHYLLACEAE

Polycarpaea sp. [not illustrated]

Low-growing, multiple stemmed, annual herb, somewhat hirsute all over; leaves small, paired. Not seen since 1990.

Frodin UPNG 8194.

Maytenus emarginata (Willd.) Ding Hou

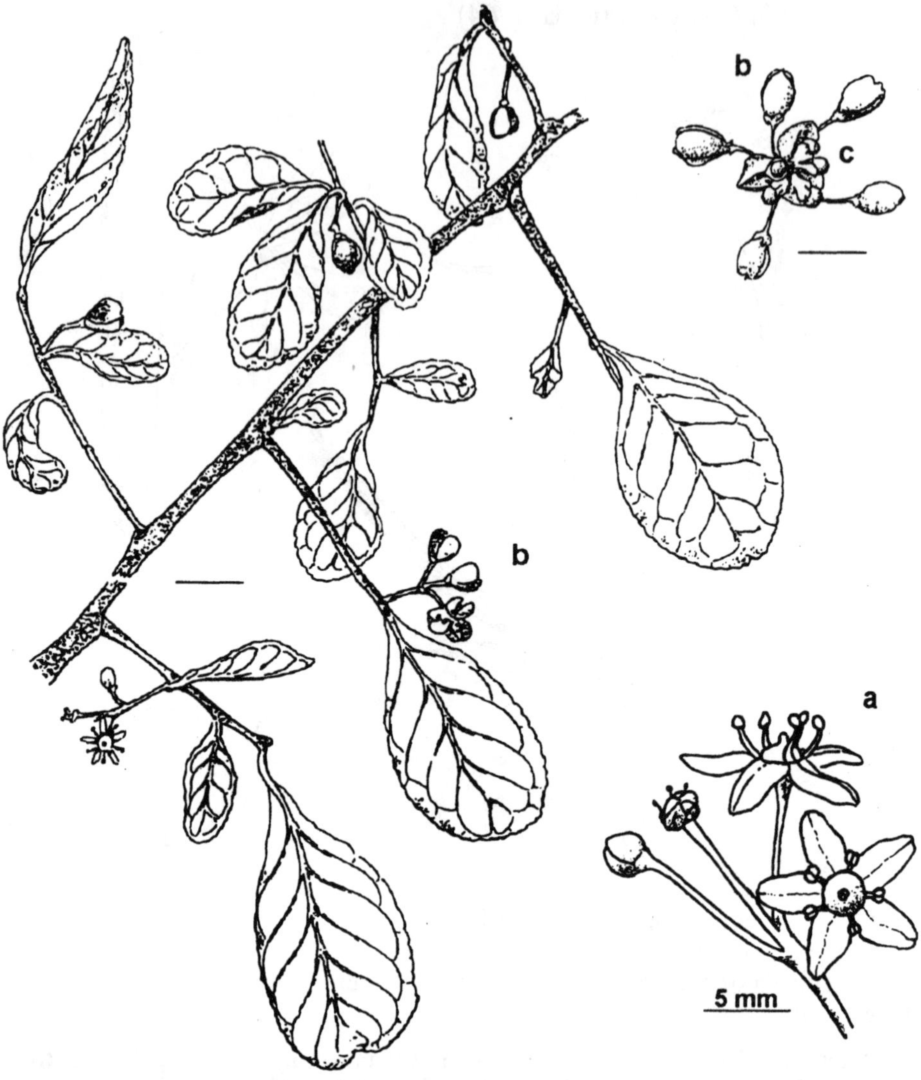

Scrambling, much branched shrub. Stems brown with short lateral shoots at right angles, each with 2 or 3 large and 2 or 3 small leaves. Leaves pale to mid-green, fairly glossy above, paler below, glabrous; margin minutely dentate. Flowers with white petals and bright yellow anthers (a). Fruit (b) inflated with 3 lines of dehiscence, green, irregularly ripening red. Seeds 1-6 per fruit, bright red-brown, part covered by white aril (c).

Behind the beach and in thicket on hill slope, common.

Sri Lanka, SE Asia, Malesia to N Queensland.

Frodin UPNG 310, 3168, 4264.

Lumnitzera racemosa Willd. [not illustrated]

(HFPNG **1**: 50; Frodin et al: 46; Fl. Mal. **4**: 588)

Small tree, 3-4 m high. Leaves oblanceolate or obovate, 3-6 x 1-2 cm, notched at apex, bright green on both sides, somewhat thick and fleshy. Flowers in small clusters in leaf axils, white. Fruit narrowly ovoid, c. 12 x 4 mm with persistant calyx, green tinged with red.

Mangrove, NE corner, on sand.

Tropical E Africa across Indian Ocean, Malesia, N Australia and Polynesia.

Frodin UPNG 5241; HCFH 1093/13625.

Terminalia microcarpa Decne. subsp. **incana** Coode [not illustrated]

(HFPNG **1**: 92; Fl. Mal. **4**: 562)

Pagoda-shaped tree, 10-15 m high. Leaves loosely clustered or scattered on ± slender twigs, grey-green above, shiny, pale yellowish beneath, with a persistant pale, felty indumentum; crown has grey-green or silvery cast. Flowers c. 3 mm long, calyx lobes ± reflexed and glabrous; ovary hairy. Fruit flattened-ovoid, 14 x 9 mm with adpressed silky hairs, fleshy, red.

Rare on Motupore but characteristic of savanna woodlands around Port Moresby. Only 2 trees recorded by Frodin on E slope beyond houses at grassland-scrub border, 5-10 m ASL. Not found in 1990.

Terminalia microcarpa is known from Malesia (excluding Malay Peninsula and Sumatra); subspecies *incana* is endemic to New Guinea.

Frodin et al. UPNG 2925, 3192.

Terminalia catappa L.

(HFPNG **1**: 72; Fl. Mal. **4**: 566)

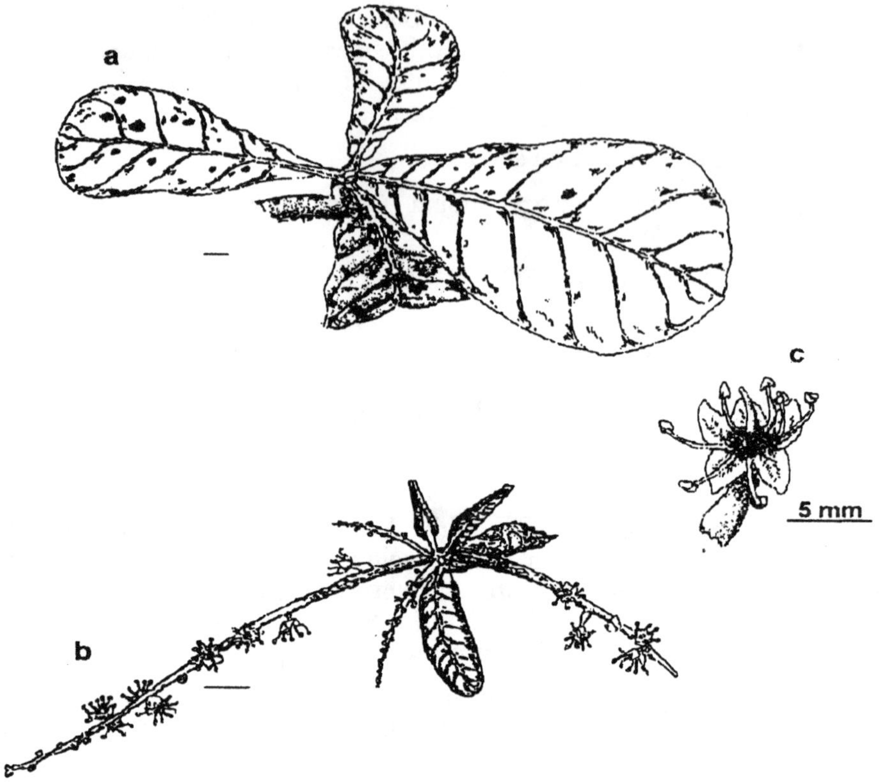

Deciduous tree, to c. 8 m high with pagoda-branching, producing 5 ± horizontal branches per layer. Leaves in clusters towards ends of thickened twigs (a) and leaving prominent leaf scars; blade mid-green, turning red when old, glabrous, slightly glossy. Inflorescence an axillary raceme at end of twig (b). Flowers (c) pentamerous; calyx lobes 5, white; corolla absent; nectaries 5, brown, hairy; producing an overpowering smell of honey. Fruit 5.5 x 3-4 cm, ± ellipsoid, somewhat flattened, surrounded by a stiff flange, somewhat fleshy with a hard fibrous stone inside and a single seed.

Strand behind mangroves on E side and planted at N end on flat grassy area.

Widespread coastal tree in Asia and Polynesia, introduced elsewhere.

HCFH 1751/13683.

Bidens sp. [not illustrated]

Herb. Leaves opposite, pinnate, the terminal leaflet deeply 3-lobed; leaflets with serrate margins and scattered hairs. Capitula on peduncles c. 10 cm long, arising towards top of plant; involucral bracts striped yellow and green; flowers yellow.

Open patch in savanna along ridge top.

HCFH 1791/13743.

Eclipta prostrata (L.)L. [not illustrated]

(Henty & Pritchard: 73)

Perennial herb. Leaves opposite, simple, lanceolate with conspicuous white hairs on both surfaces and on stems; margin shallowly serrate. Capitula 2 or 3 together in upper axils; peduncles 2-5 cm; flowers white or greenish.

Station area and along strand behind mangroves.

Pantropical weed.

HCFH 1767/13699; Kembi s.n. 22/7/75.

Emilia sonchifolia (L.)DC. [not illustrated]

(Henty & Pritchard: 76)

Erect herb. Leaves alternate, thin, lower ones stalked, upper ones sessile and clasping stem, with a few scattered hairs. Capitula in a loosely branched, terminal inflorescence; involucral bracts ± covering the flowers which are white at the base and pinkish-violet at apex; bracts strongly reflexed when in fruit.

Station area and weed in flower beds.

Pantropical weed.

Frodin & Pernetta UPNG 6653; HCFH 1768/13700.

Sonchus oleraceus L. [not illustrated]

(Henty & Pritchard: 83)

Erect, single-stemmed annual herb to 1 m. Stem and leaves somewhat succulent; sap milky. Leaves irregularly lobed and dentate, the upper sessile, clasping the stem. Capitula in a loose terminal panicle. Corolla yellow, seeds plumed.

Only seen once, on north summit.

Eurasian origin, now world wide.

JM 95/140.

Synedrella nodiflora (L.) Gaertn. [not illustrated]

(Henty & Pritchard: 86)

Erect herb. Leaves opposite, simple, petiolate, with scattered white hairs on both sides; margin shallowly serrate. Capitula sessile, clustered in leaf axils, yellow.

Station area.

Pantropical weed of American origin.

Frodin & Pernetta UPNG 6649; HCFH 17701/13702; Lawong s.n. 30/8/88.

Tridax procumbens L. [not illustrated]

(Henty & Pritchard: 87)

Decumbent herb. Leaves opposite, covered with long white hairs giving a hoary appearance; margin irregularly serrate to lobed. Capitula solitary, on terminal peduncle 20 cm long; flowers pale yellow.

Station area.

Tropical weed of American origin.

Frodin & Pernetta UPNG 5257, 6673; HCFH 921/13452, 1772/13704.

Vernonia cinerea (L.) Less. [not illustrated]

(Henty & Pritchard: 88)

Erect herb. Stems ribbed. Leaves alternate, stalked, with very shallowly toothed margins and a few short scattered hairs. Capitula small, in terminal panicles; corolla white with purple lobes.

Station area and savanna along nature trail, on ridge top and slopes.

Palaeotropical weed, introduced into S America.

Frodin s.n. 1972; HCFH 629/13160, 1769/1370; Lawong s.n. 24/8/88.

Evolvulus alsinoides (L.)

(Fl. Mal. **4**: 395)

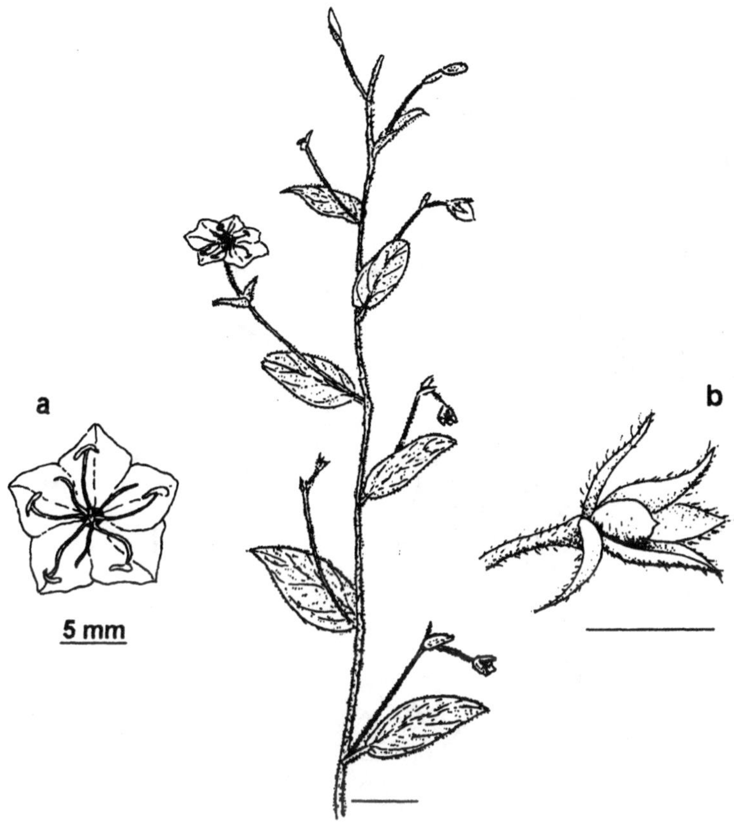

Upright, much branched slender-stemmed herb, entirely covered with soft white hairs. Leaves ovate, ± sessile. Flowers solitary in the axils (a); corolla pale blue-mauve. Fruit (b) a green, globose capsule.

Occasional, among grass on ridge top and south west slopes.

Tropical E Africa, S Asia, Malesia, except Borneo, to Australia.

JM 93/55.

Ipomoea carnea Jacq. [not illustrated] is an ornamental shrub: leaves alternate, petiolate, cordate with attenuate apex; flowers in terminal inflorescence; corolla tubular, mauve. Behind houses, Station area. Adventive or cultivated; not seen in 1990. Native of tropical America. *Huxley s.n. 15/6/75.*

Ipomoea macrantha Roem. & Schult.

Moon Vine.

(Fl. Mal. 4: 487)

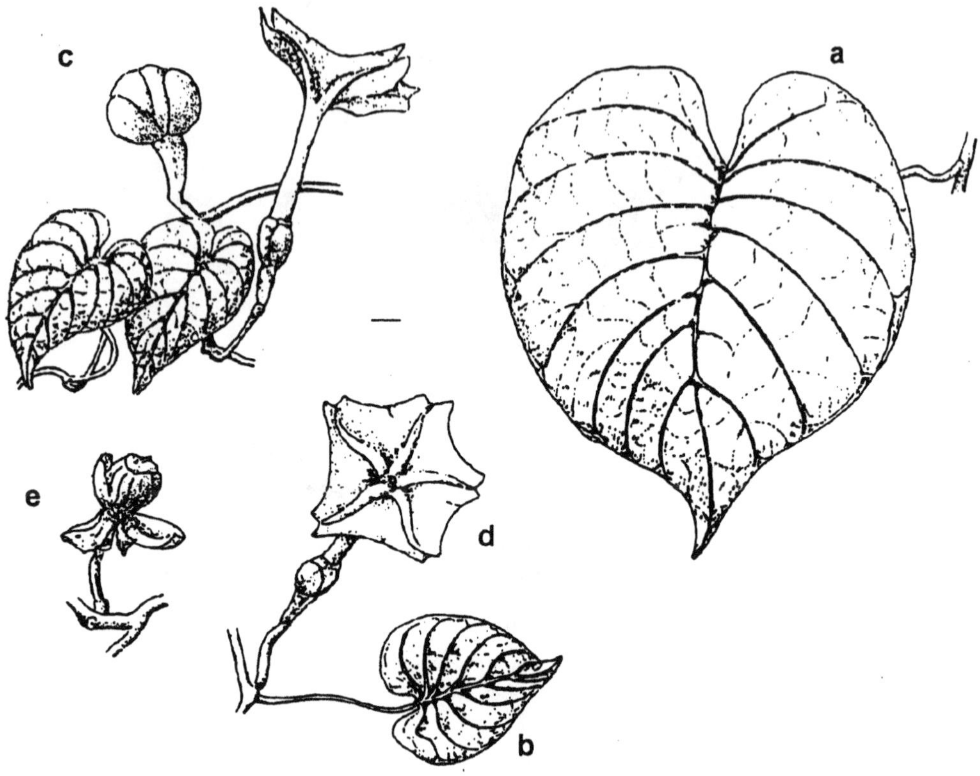

Twining climber to 5m high. Stems green-brown. Leaves alternate, mid-green, glabrous, rather dull, very variable in size, the lower ones (a) much larger than the upper (b). Flowers in small groups in upper leaf axils; buds straw-coloured (c). Corolla long, trumpet-shaped, white, nocturnal (d), withering by early a.m.. Fruit a dry capsule, 3 cm diameter, splitting at the top into 4 x 1-seeded sections (e). Young fruit covered by whitish-green sepals; older ones brown, exposed. Seeds bearing 2 rows of hairs.

NW corner, back of mangroves, in full sun.

Tropical America, E Africa, Indian Ocean, continental Asia to Polynesia.

Frodin UPNG 801; HCFH 925/13456, 1086/13618.

Ipomoea nil (L.) Roth
(Fl. Mal. **4**: 465)

Twining creeper. Leaves alternate, far-spaced on stem, 3-lobed, yellow-green above, paler below, texture soft, slightly hispid. Flowers opening early a.m., closing soon, solitary or clustered on peduncle in leaf axil. Corolla tube white, lobes blue, fading to purple. Fruit dry, brown.

Abundant in 1993, especially on forest margin, west side but not seen in 1990-92.

Circumtropical.

Frodin UPNG 4251; Huxley s.n. 15/6/75; JM 93/68.

Ipomoea pes-caprae (L.) Sweet
(Fl. Mal. **4**: 475)

Prostrate creeper. Stems purplish. Leaves alternate, petiolate, somewhat succulent, broadly elliptical to round with notched apex, glabrous. Flowers axillary in 1- to few-flowered clusters. Corolla trumpet-shaped, purple-red (a). Fruit a dry capsule, 2.5 cm diameter, splitting at the top into 4 sections when mature.

On beach at NE corner of island and on rocks to 1 m ASL, on west side.

Circumtropical.

Frodin UPNG 5243; Yanape MY 24.

Jacquemontia paniculata (Burm. f.) Hallier f.
(Fl. Mal. **4**: 432)

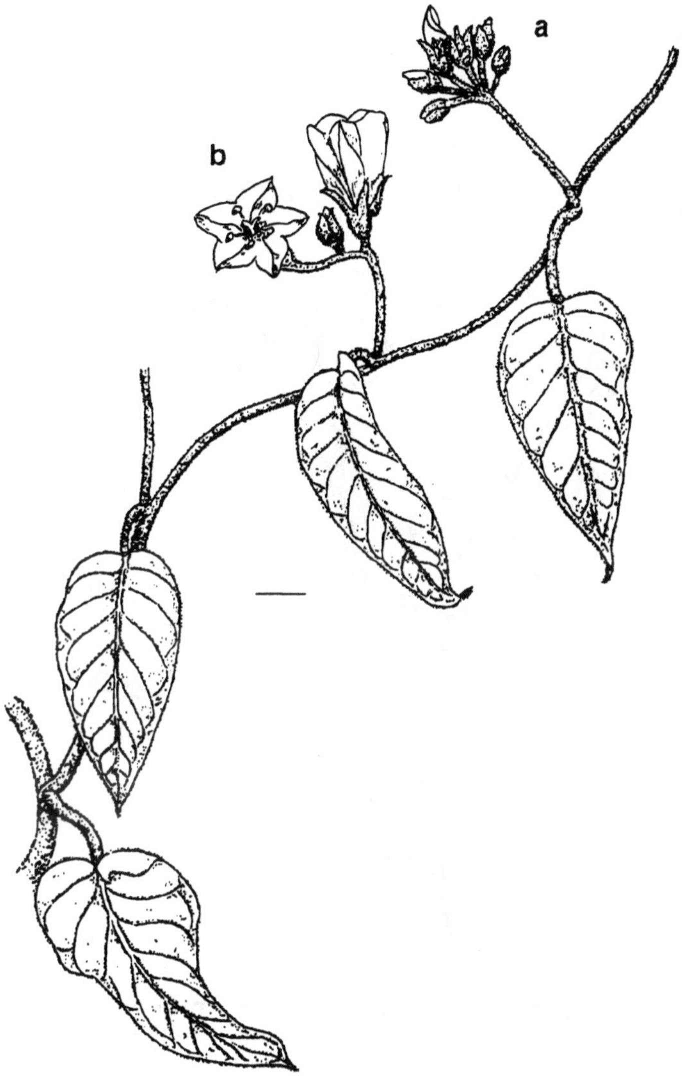

Herbaceous twiner. Leaves alternate, cordate, yellow-green or pale green above, paler below, softly hairy, mucronate. Flowers c. 6-10 in axillary clusters (a). Corolla tubular, mauve (b). Fruit a small capsule, c. 0.5 cm diameter with persistant calyx; fruit and calyces light green with 4 or 8 valves.

In grassland and edge of coastal scrub, E & W sides of island.

Africa, Asia to Australia and New Caledonia.

Frodin UPNG 3172, 4248, 4251A, 5251; Huxley s.n. 15/6/76; Kuduk & Naoni UPNG 6113; Powell UPNG 2326.

Merremia tridentata (L.) Hallier f.

(Fl. Mal. **4**: 445)

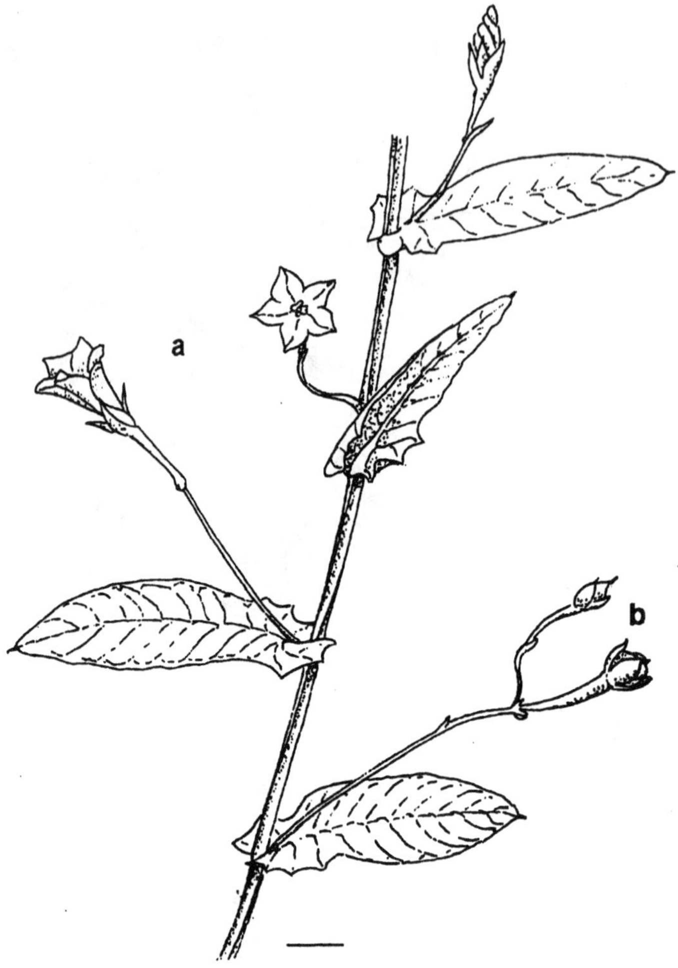

Twining herb to 3 m; stems winged; leaves pale green, glabrous, almost sessile, clasping; flowers 1 or 2 per axil, on long stalks; calyx green, 5 broad teeth; corolla tubular, white, 2 cm diameter (a). Fruit a globose capsule (b).

Common at edges of scrub patches on W slopes.

East Africa through India and Malesia to China and Australia.

JM 93/67.

Frodin UPNG 8216, collected in 1984, represents an additional, unidentified species of Convolvulaceae. This plant is similar to *Merremia tridentata* but the leaves are long-petiolate.

CUCURBITACEAE

Diplocyclos palmatus (L.) C. Jeffrey

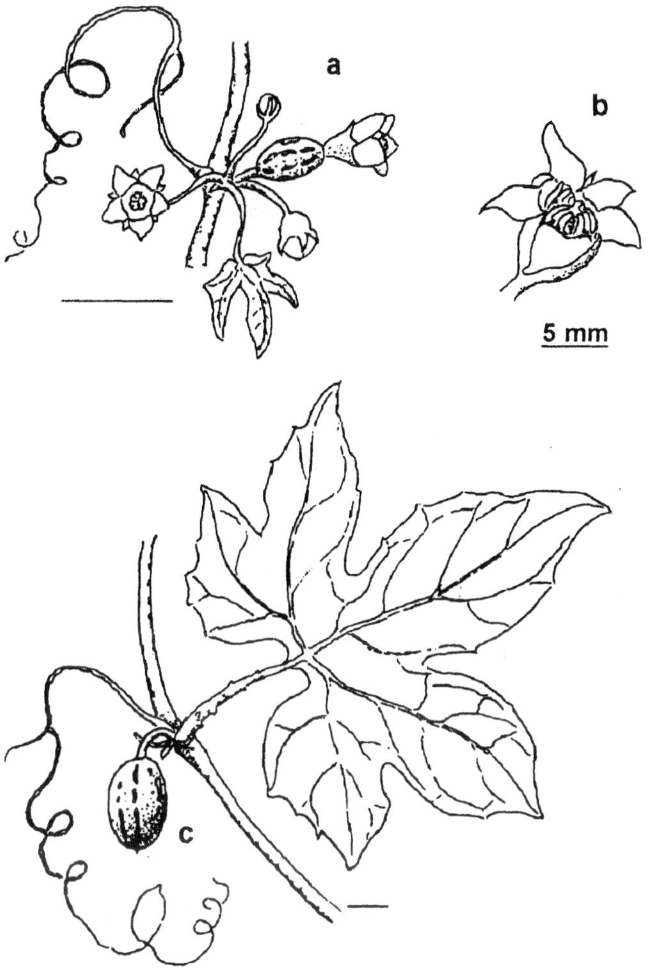

Herbaceous climber to 3-4 m high. Stems grooved. Leaves alternate, petiolate, deeply 3-lobed, the lobes sometimes lobed themselves; upper surface covered with white dots, which are the swollen raised bases of short hairs. Branched tendril opposite each petiole. Monoecious; several male and one female flower at each axil (a); flowering only near growing tips; corolla yellow; male flower (b). Fruit ovoid, green ripening orange, striped white (c); seeds blackish, in translucent pulp.

Savanna and scrub margin, W side.

Tropical Africa, Asia, Malesia, Western Pacific and Australia.

Frodin UPNG 5267; JM 93/65; Kuduk & Naoni UPNG 6109.

Momordica charantia L.

(Henty & Pritchard: 98)

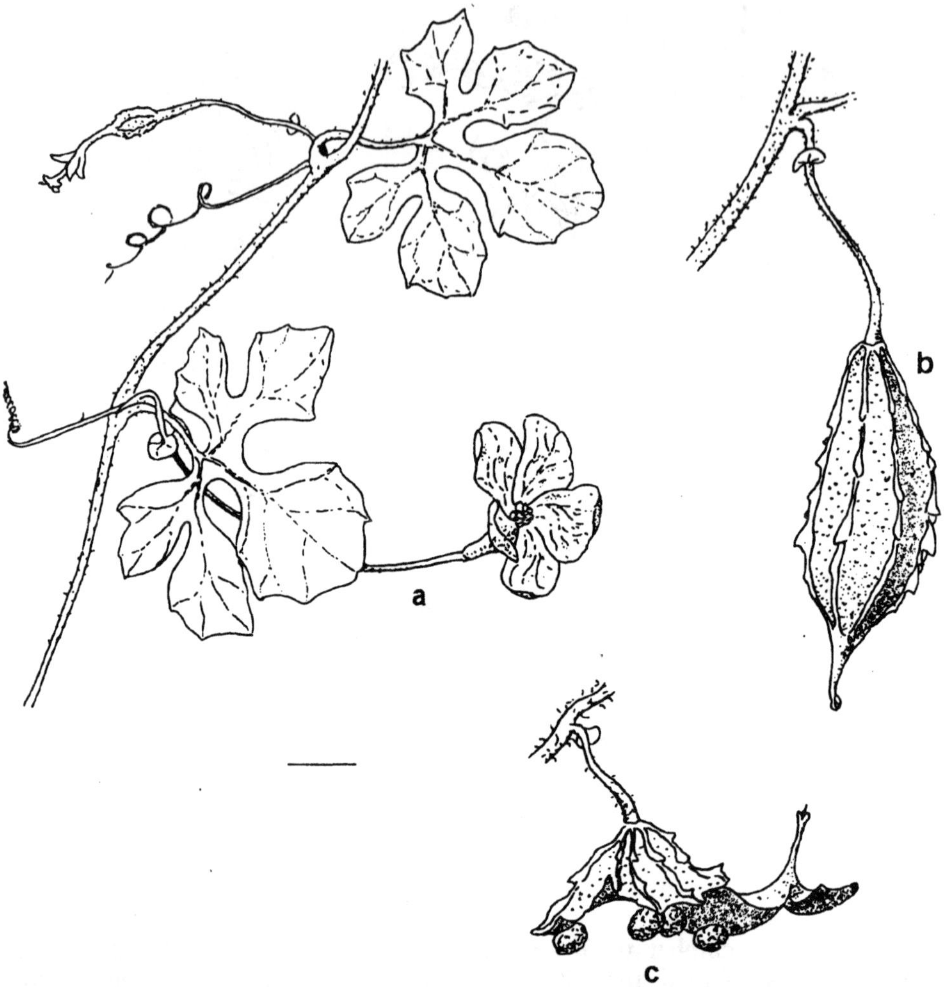

Herbaceous trailing climber. Stems grooved when dry. Leaves alternate, petiolate, palmatifid, deeply 5-7 lobed, the lobes rounded or obovate; upper surface with scattered white hairs; lower surface covered with white dots; margin toothed, the teeth mucronate. Unbranched tendril opposite each petiole. Monoecious. Flowers solitary in axils; corolla yellow, deeply 5-lobed (a). Fruit pendulous, warty, orange (b), breaking open to reveal black seeds embedded in red pulp (c).

Behind houses, Station area, at edge of tall grass.

Pantropical; probably introduced into America.

Frodin UPNG 3161.

Mukia maderaspatana (L.) M.Roemer

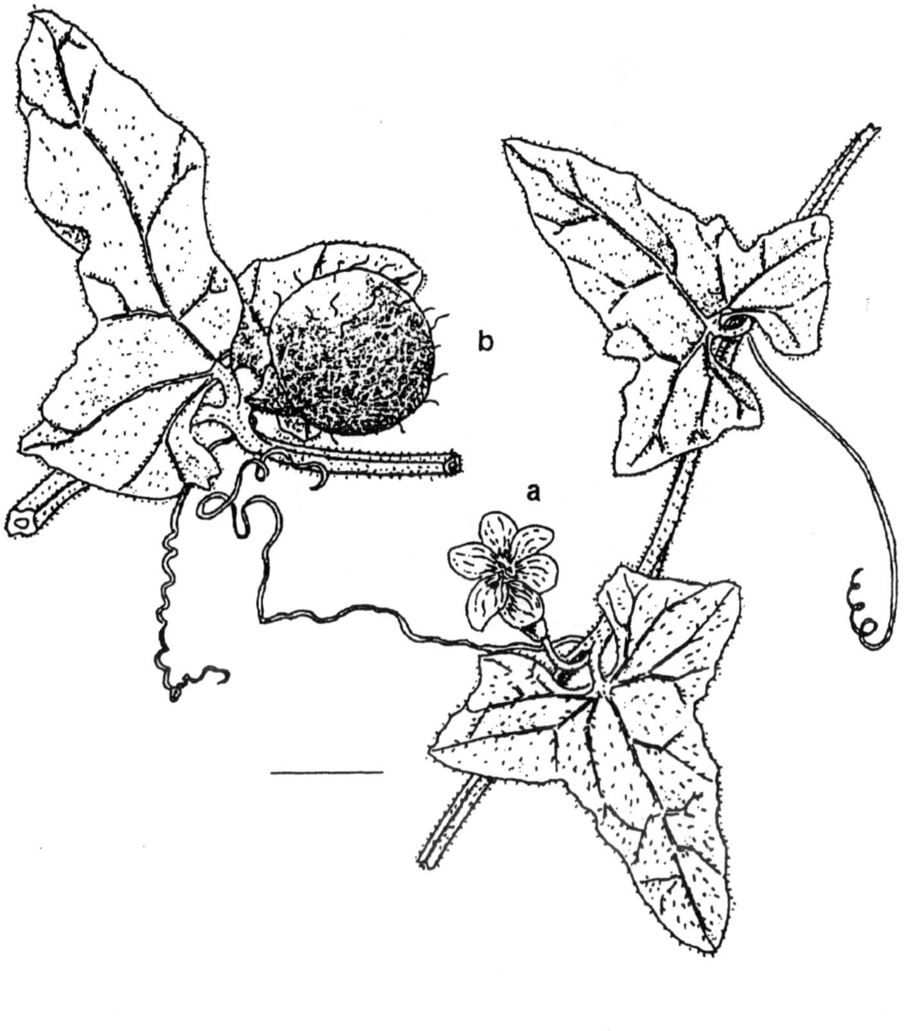

Herbaceous climber. Stems square, drying grooved and sometimes striped Leaves alternate, petiolate or ± sessile, shallowly 3-lobed to somewhat triangular, cordate at base; dull green above, paler below; margin sinuate-aculeate. Fine white hairs on both surfaces, texture rough. Unbranched tendril opposite each petiole. Monoecious. Flowers small, in axillary clusters, yellow (a). Immature fruit green streaked with white. Mature fruit red, fleshy (b), containing numerous seeds.

Frequent in savanna along ridge top and elsewhere on slopes.

Africa, Asia, Malesia to tropical Australia.

Frodin UPNG 4282, s.n. 1972; HCFH 1080/13612, 1081/13613, 1734/13666; Huxley s.n. 15/6/75; Kuduk & Naoni UPNG 6111.

Zehneria sp.

Delicate climbing herb, to 1-2 m high; stems terete, glabrous. Leaves deeply 3- to 5-lobed, pale green, rather flaccid, minutely hispid; tendril unbranched. Flowers axillary, 1 male and 1 female per axil, on long pedicels (a); corolla white. Fruit elongate, to 2 cm, green, ripening translucent pink (b).

Frequent in grass and at thicket margins, W side near strand.

JM 93/64.

There is no exact match for this plant at the Kew herbarium but the fruits resemble *Z. japonica* (Thunb.) H.Y. Lui (C. Jeffrey, pers. comm.).

EBENACEAE

Diospyros littorea (R.Br.) Kosterm. [not illustrated]

D. ferrea (Willd.) Bakh. var. *littorea* (R.Br.) Bakh. pp.

Tree to 8 m high. Leaves alternate, glabrous; petiole short, curved; lamina narrowly elliptical, glabrous, dark green above and much paler below; margin entire. Dioecious. Flowers small, axillary; corolla c. 6 mm long. Fruit an oval-elliptical berry c. 14 x 10 mm, with a persistant calyx, the lobes adpressed to the fruit; seeds 5-6, wedge-shaped.

Scrub on E side, beyond Caretaker's house; not found in 1990.

A segregate of the widespread palaeotropical *D. ferrea*, recorded from Thailand, New Guinea and N Australia.

Frodin UPNG 5215.

ELAEOCARPACEAE

Elaeocarpus arnhemicus F. Muell.

(HFPNG **2**: 93)

Spreading tree 2-15 m high. Twigs bearing prominent pale lenticels (a). Leaves alternate, glabrous, glossy, mid-green above, paler below; margin crenate; paired glands on petiole at base of lamina, and glands (domatia ?) sometimes on underside at junction of secondary veins with main vein. Flowers in pendent axillary racemes (b). Sepals white; petals white and fringed; nectary orange; stamens buff; strongly smelling of honey (c). Fruit a drupe (d, e), usually one per inflorescence, green turning deep blue; flesh green, containing a single, very hard, sculptured stone.

Common in scrub, E & W sides of island, especially near strand line.

Northern Australia and New Guinea; restricted to areas with a distinct dry season.

Frodin UPNG 1367, 4272; HCFH 1058/13590, 1045/13577, 1070/13609; Powell et al. UPNG 2323.

Elaeocarpus arnhemicus F. Muell.

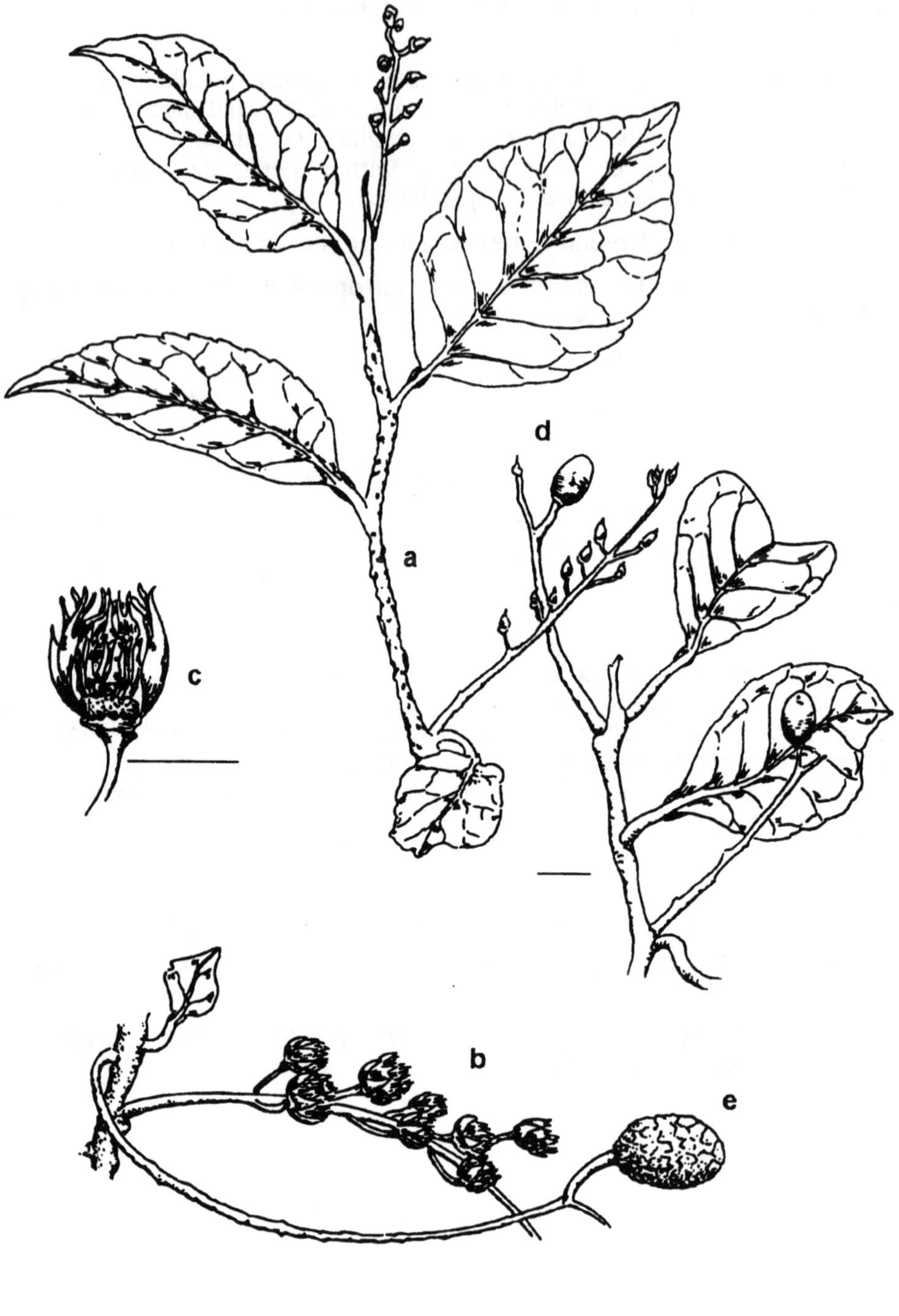

Breynia cernua (Poir.) Müll. Arg.

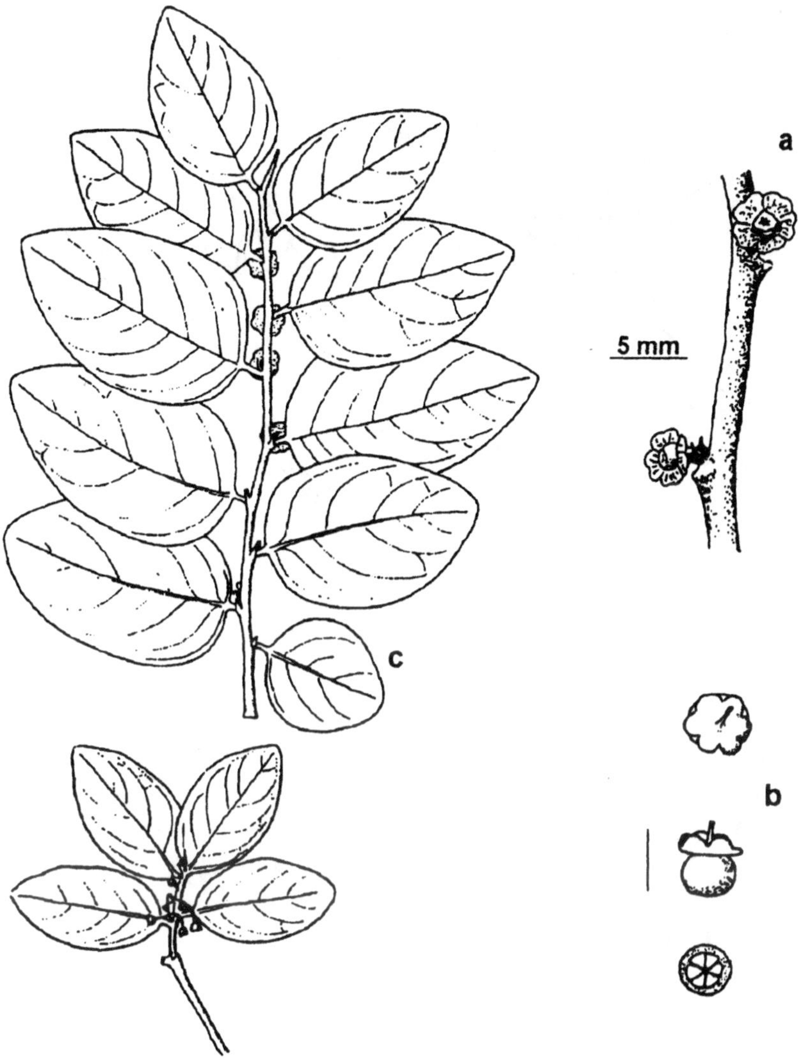

Shrub or small tree to 3 m high. Leaves alternate, distichous, glabrous, very thin, flacid, medium to pale green; venation slightly prominent; margin entire. Flowers unisexual, very small, one to several per leaf axil; female (a) yellowish, fleshy. Fruit (b) a pink berry, spherical, with a 6-lobed cream "frill" (persistant calyx); flesh containing 6 black seeds in 3 groups of 2; 1 fruit per leaf axil, as in (c).

Savanna, on ridge top.

Java, Philippines, E Malesia to N Australia and Solomons.

HCFH 1041A/13573; 2018/13770.

Bridelia oligantha Airy Shaw

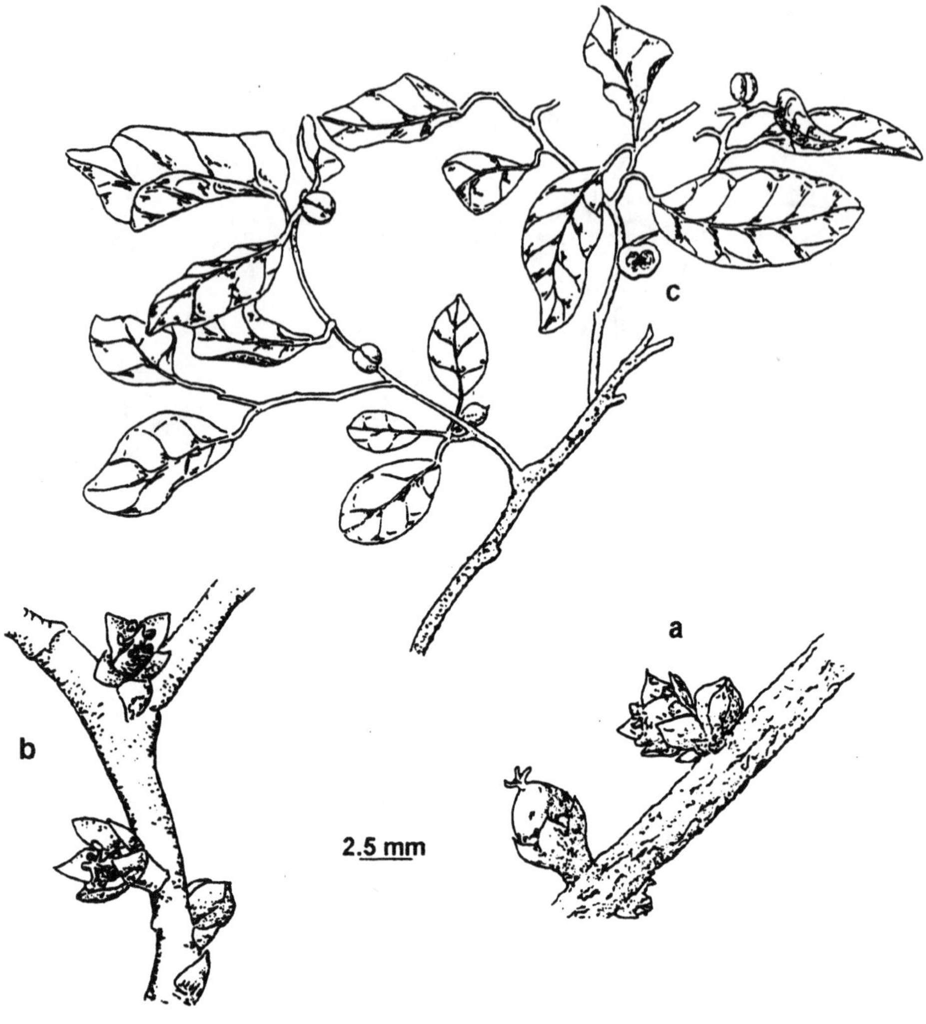

Small spreading tree or shrub. Twigs brown with prominent lenticels. Leaves alternate; petioles twisted; lamina dull yellowish green above, greyer below, glabrous; margin slightly sinuate. Flowers unisexual, yellow; female (a), male (b). Fruit (c) smooth, translucent, 2-seeded, fleshy berry, greenish turning black when fully ripe; containing 2 seeds (c).

Frequent in savanna, scrub edge and near shore.

Endemic to New Guinea.

Frodin UPNG 844, 850, 4246, 5226; HCFH 1056/13588, 1738/13670.

The generic name is also spelt *Briedelia*.

Drypetes australasica (Müll. Arg.) Pax & Hoffm.

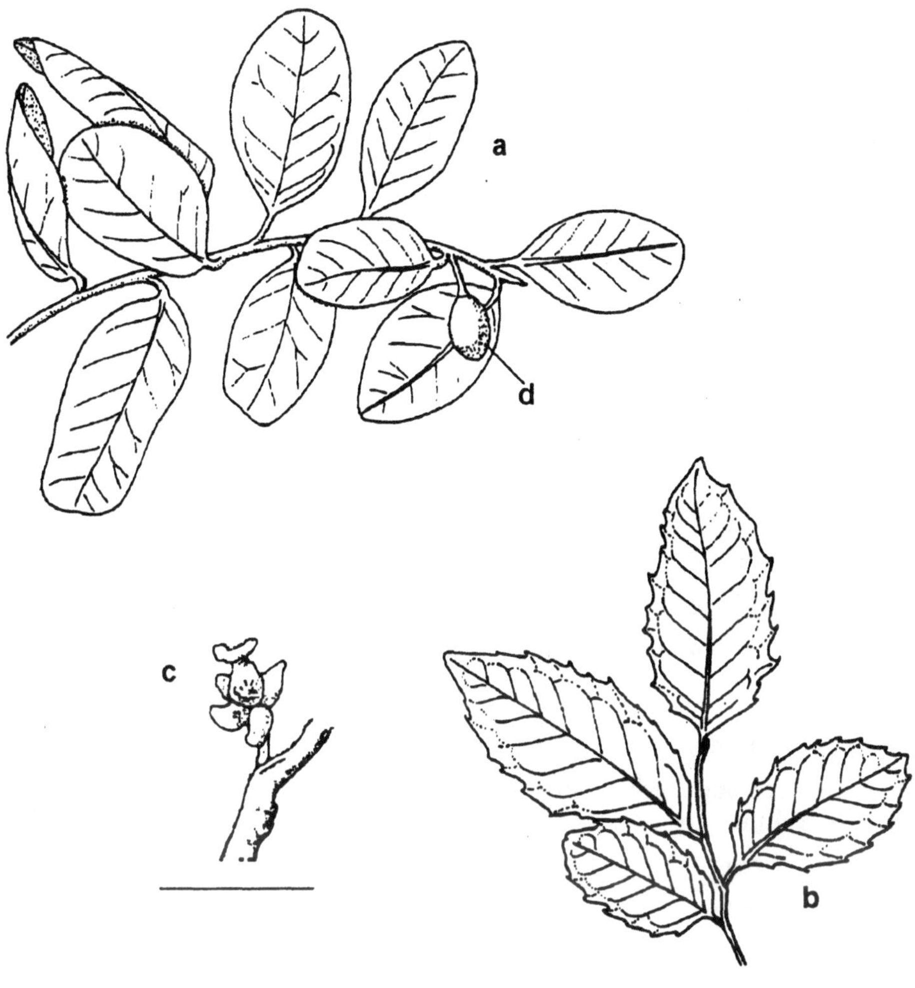

Small bushy tree 4-5 (-11) m high. Twigs grey with paler lenticels. Leaves spirally arranged, thick, ± leathery, stiff, glabrous, dark glossy or dull green above, only slightly paler below, slightly revolute; margin entire (a); venation visible above, less so below. (*HCFH 2000/13754* may be a variant with aculeate leaf margins) (b). Dioecious. Flowers minute; buds white; female flowers (c) yellowish with 4 petals and large stigma. Fruit (d), on a short pedicel; exocarp orange-red, smooth, containing a single seed.

Frequent in scrub on E & W sides of island and strand on W side.

Northern Australia and New Guinea.

Frodin UPNG 842, 853, 4265, 4291A, 4488; HCFH 614/13145, 1044/13576, 1064/13596, 1097/13629; Huxley s.n. 15/6/75; Kuduk & Naoni UPNG 6121; Powell et al. UPNG 2345.

Euphorbia heterophylla L. [not illustrated]

E. geniculata Ort

(Henty & Pritchard: 102-103)

Erect annual herb to 50 cm high, with milky sap. Leaves alternate, ovate. Flowers unisexual, much reduced, in cyathia grouped in terminal inflorescences. Cyathium a 5-lobed green cup. Fruit 3-lobed, with 1 seed per lobe.

Station area; weed. American origin, now pantropical.

HCFH 1786/13718.

Euphorbia hirta L. [not illustrated]

(Henty & Pritchard: 104)

Ascending herb with milky sap. Stems densely covered with yellow hairs. Leaves opposite, small, obliquely rhombic-elliptic, dull dark green above, grey-green below, sometimes blotched. Cyathia in dense axillary globose inflorescences.

Station area.

Pantropical weed of S American origin.

Frodin & Pernetta UPNG 6665; HCFH 1773/13705, 1774/13706.

Euphorbia prostrata Ait. [not illustrated]

Low creeping herb. Stems reddish, glaucous. Leaves opposite, very small, c. 6 x 4 mm, ovate-obovate, reddish green. Cyathia in leaf axils.

Station area.

Widespread tropical weed, originally from Jamaica.

Frodin & Pernetta UPNG 6646.

Euphorbia vachellii Hook. & Arn. [not illustrated]

E. serrulata Reinuv. ex Bl.

Stiff erect herb, to 50 cm high. Lower stems reddish. Leaves opposite, linear, with a few scattered hairs, yellow-green above; margin serrulate. Cyathia in dense terminal pseudocymes; bracts white; flowers tiny. Fruit green.

Savanna, along ridge top and nature trail.

E Asia, Malesia to N Australia, Solomons, Vanuatu.

Frodin UPNG 4287; HCFH 630/13161, 1746/13678.

Euphorbia plumeroides Teijsm. ex Hassk.

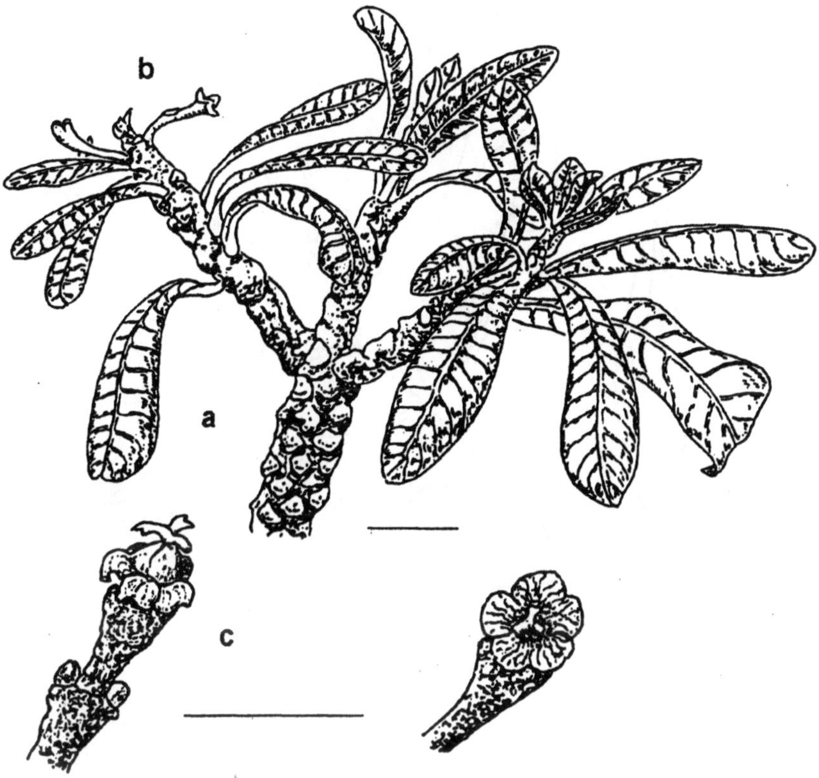

Shrub to 1.5 m high, much branched, dichotomizing; stunted appearance. Stems thick, rough, with prominent leaf scars (a). Sap milky. Leaves in clusters at ends of twigs, dull dark or olive-green, glabrous; margin entire. Cyathia greenish, 5-lobed (b & c) in subterminal pairs. Flowers minute; ovary 3-sided with 3 stigma lobes (c).

Rare; several clumps on rocky promontaries on W side, 10-15 m ASL, savanna.

Indonesia, Philippines, New Guinea, northern Australia and Fiji (?introduced).

Frodin UPNG 2940, 3174; HCFH 1731/13663.

Excoecaria agallocha L.

(Henty: 52 & 59)

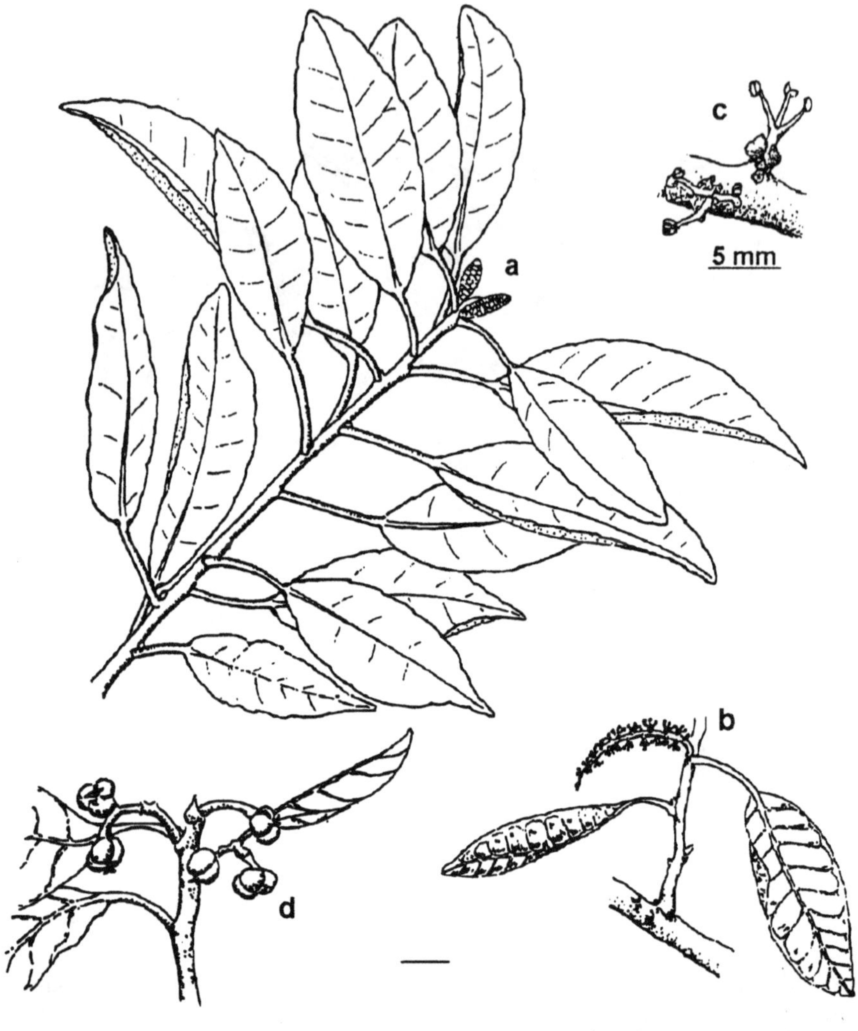

Bushy tree, to 5+ m high. Exudes milky sap. Twigs smooth, brown, glabrous. Leaves alternate, glabrous, dark or mid-green above, paler below; venation ± obscure; margin crenulate. Petiole yellowish, with 2 glands on upper surface at junction with blade. Dioecious. Flowers small. Young male inflorescence yellow-green, cone-like (a), at apex of twigs or in leaf axils, later elongating (b); flowers minute, with 3 tepals and 3 stamens (c). Female inflorescence branched. Fruits 3-lobed, smooth, green (d).

Behind mangroves on E side and on rocky shore on W side.

S India to E Asia, Malesia, N Australia and Pacific; edge of mangroves and fresh water swamps.

Frodin UPNG 805, 811, 813; HCFH 662/13153, 1049/13581, 1094/13626.

Flueggea virosa (Roxb. ex Willd.) Voigt

Securinega melanthesoides (F. Muell.) Airy Shaw

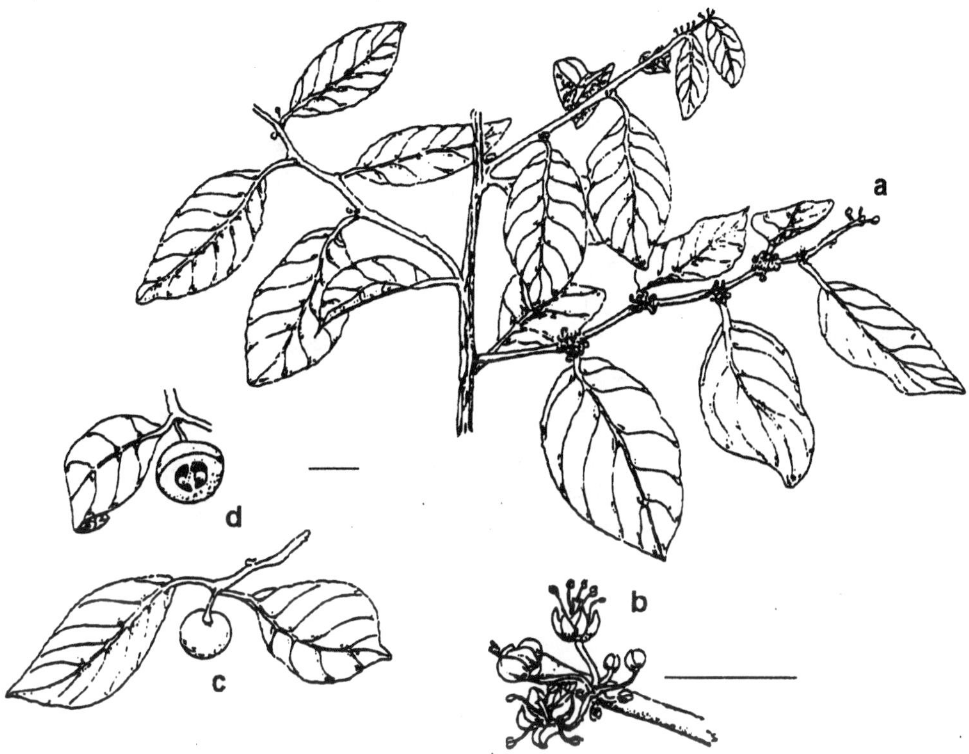

Spreading shrub to 3 m high. Stems brown, grooved. Leaves alternate, dull mid-green or grey-green above, often a bit mottled, glaucous below, glabrous; margin slightly crenulate. Flowers tiny in ball-like clusters at nodes (a & b), yellowish or greenish white. Fruit a pure white fleshy berry (c), containing 2 brown seeds (d).

Savanna, especially near north end.

New Guinea and Australia.

Frodin UPNG 4267, 4284; HCFH 626/13157, 1034/13586; Huxley s.n. 15/6/75; Lambinon 87/108; Tippett UPNG 828.

Mallotus philippensis (Lam.) Müll. Arg.

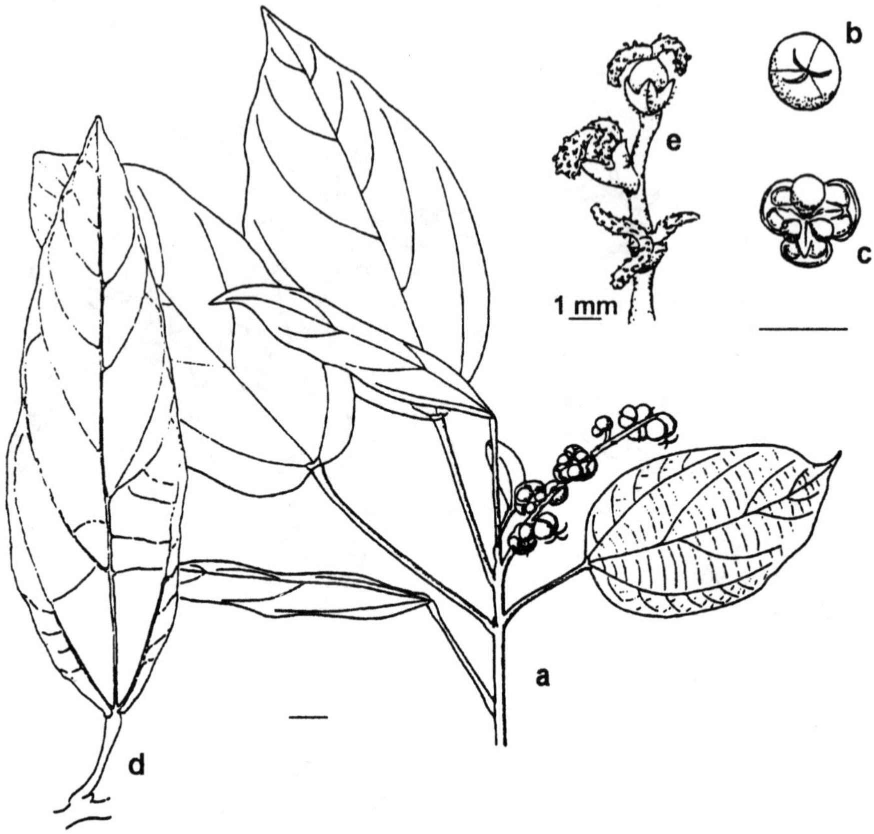

Shrub or small tree. Stems and petioles brown, covered in brown pubescence. Leaves (a) alternate, dull mid-green, ± glabrous above, paler or glaucous and ± felty below; venation yellowish, raised on upper surface; margin ± entire, slightly undulate. Petioles angled at top so blade is ± perpendicular to twig; 2 glands at junction of petiole and blade on upper surface. Dioecious. Female flowers small, inconspicuous, brown, in spikes in upper leaf axils or terminal. Fruit a capsule with 3 persistant styles (b), covered in dull red, scurfy, dust which rubs off, and containing 3 black shiny seeds (c).

Rare, several together in scrub on E side on slope above Caretaker's house. A larger tree (*JM 93/66*), on the ridge at S end, has more elongate leaves and shorter petioles (d) and may represent a different species. Female flowers (e) were drawn from this tree.

W Himalayas and Sri Lanka, S & E Asia, Malesia to Australia and Melanesia.

Frodin UPNG 3181; HCFH 1711/13664, 2028/13780; JM 93/66.

Manihot esculenta Crantz [not illustrated]

Tapioca or cassava.

(Henty: 61)

Woody herb, shrub or small tree to 2 m high. Sap milky. Leaves alternate, simple or palmatifid with 3-9 segments; glabrous. Petioles crimson. Flowers in yellowish racemes, lower ones female. Fruit globose, with 6 narrow wings; 3-celled.

Cultivated behind houses at edge of tall grass; not seen in 1990.

Native of Brazil.

Frodin UPNG 3160.

Pedilanthus tithymaloides (L.) Poit. [not illustrated]

(Henty: 61)

Succulent herb to c. 0.5 m high. Sap milky. Leaves alternate, distichous, green, often mottled with white. Flowers red.

Cultivated near Old Hostel and escaped into grassy edge of mangrove, NE corner.

Native of tropical America. Cultivated as an ornamental.

Phyllanthus amarus Schumach. [not illustrated]

(Henty & Pritchard: 106, as *P. niruri* L.)

Small annual herb to 20 cm high. Leaves alternate, distichous, small, 9 x 4 mm, elliptical. Flowers unisexual, small, greenish. Fruit a smooth, green capsule, c. 2 mm diameter, pendant below leaves; 3-celled.

Station area.

Pantropical weed originating in America.

Frodin & Pernetta UPNG 6675; HCFH 1771/13703.

FLACOURTIACEAE

Scolopia novoguineensis Warb.

S. nitida C.T. White.

(Fl. Mal 5: 12, as *S. nitida*)

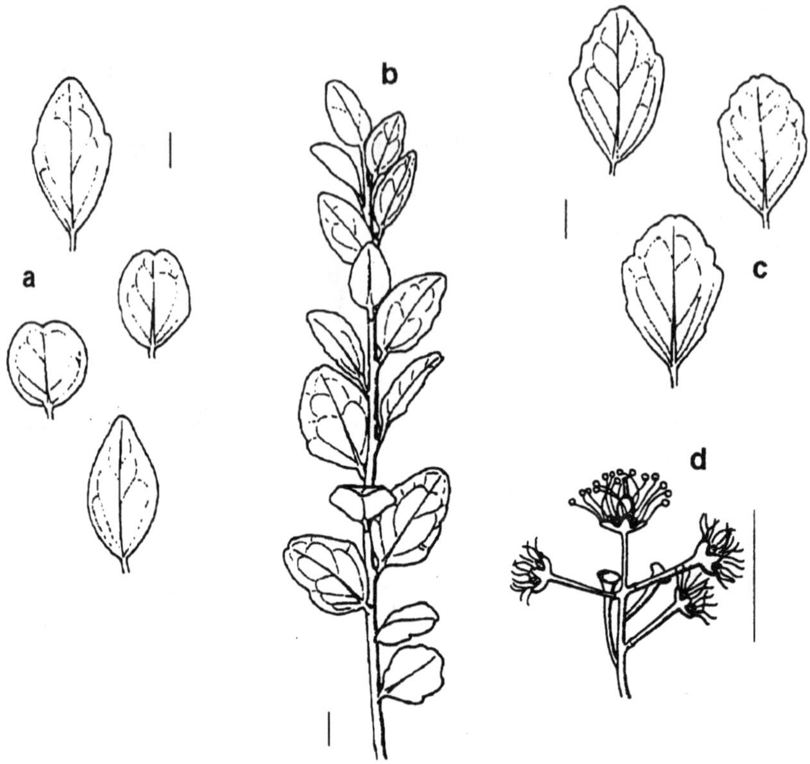

Shrub with numerous upward-pointing stems. Erect sucker shoots bearing spines up to 3 cm long, otherwise spineless. Leaves alternate, young leaves small, to 2.3 x 1.5 cm, ± rhombic, mucronulate, bright green; older leaves leathery, glabrous, shiny olive- or grey-green above, slightly or much paler beneath; margin irregularly notched (a, b & c). Flowers in short axillary racemes (d), white. Sepals and petals 4-6; stamens numerous; ovary superior. Fruit a berry, yellow turning red, with orange pulp.

Rare; just above shore line on E side of island.

New Guinea and Bismarck Archipelago.

Frodin UPNG 5246; HCFH 1764/13696 (juvenile), 2001/13753 (adult).

GOODENIACEAE

Scaevola sericea Vahl, syn. *S. taccata* Roxb. [not illustrated], is an Indo-Pacific strand plant collected in 1976 on the shore behind the mangroves, but not seen since: shrub, slightly fleshy to 1 m high, leaves ovate, clustered at branch ends; flowers white, zygomorphic, with all 5 lobes lying in lower half; berries white, fleshy. *Frodin UPNG 3188.*

Anisomeles malabarica (L.) R.Br. ex Sims

A. salvaiifolia R. Br.

(Henty & Pritchard: 107; Fl. Mal. **8**: 329)

Woody herb 0.5-1.5 m high, softly hairy throughout. Stems square. Leaves opposite, ovate, shortly petiolate; margin dentate or crenate; smell unpleasant. Flowers in dense clusters in terminal leaf axils and at apex of shoots; calyx with 5 teeth; corolla pale mauve or pale blue, the upper lip short, not covering the stamens (lateral view of flower, a).

Among grass on steep W slope and elsewhere.

Asia & Malesia to Bismarck Archipelago.

Frodin UPNG 2568; JM 91/5, 93/58, 93/62.

Hyptis suaveolens (L.) Poit.
(Henty & Pritchard: 110; Fl. Mal. **8**: 371)

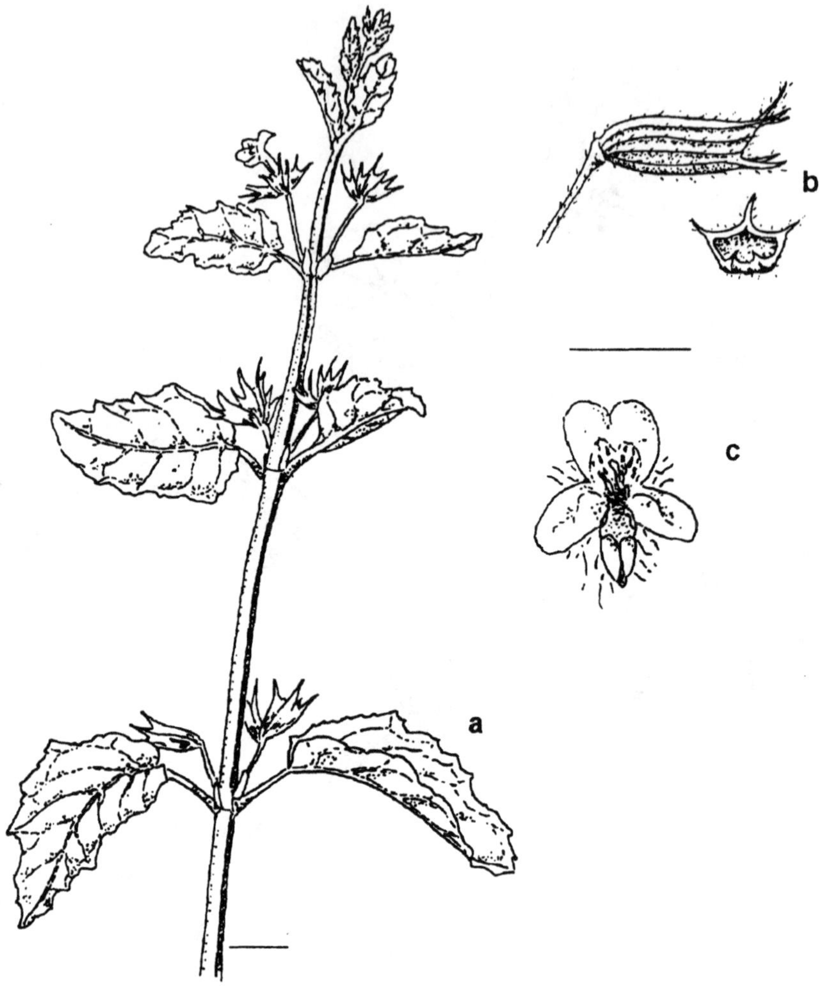

Strongly aromatic herb, to 0.5 m. Stems square. Leaves opposite, petiolate, cordate, hairy; margin irregularly dentate or doubly serrate (a). Flowers in loose, stalked, axillary clusters; calyx flattened, ridged, with 5 very thin teeth, 3 dorsal and 2 ventral (lateral and front views, b); corolla blue-violet (front view of flower, c).

In grass near houses and on slopes.

Native of tropical America; naturalized throughout tropics.

HCFH 1778/13710; JM 92/38.

Leucas flaccida R. Br.

(Fl. Mal. **8**: 337)

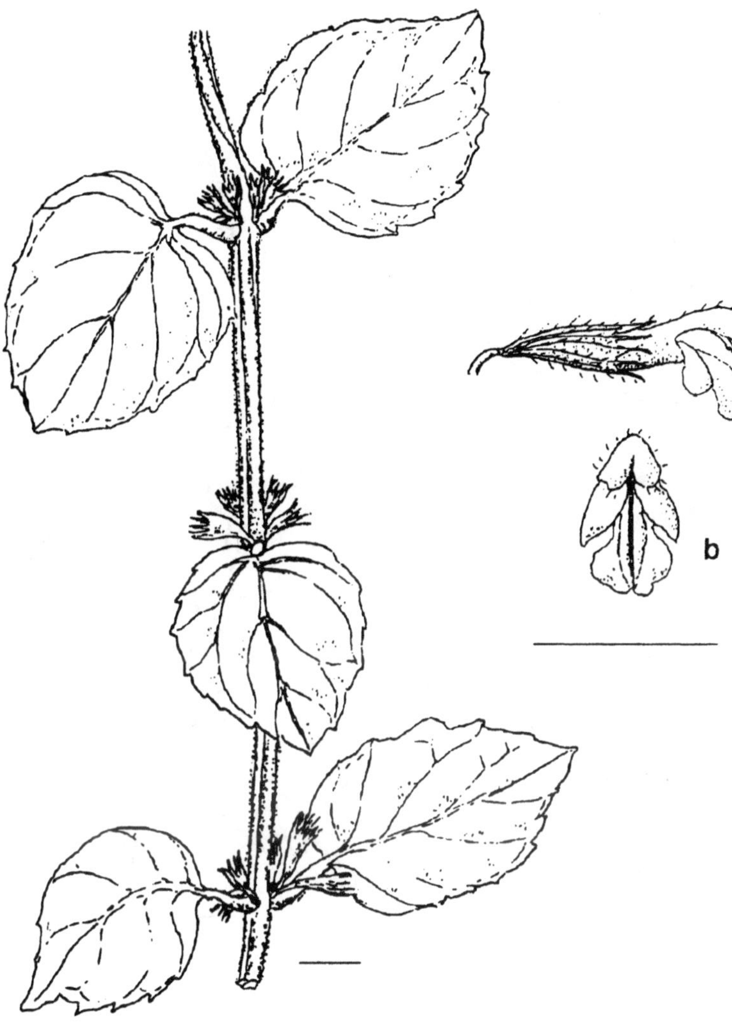

Slightly woody herb to 0.5 m; stems square, softly hirsute; leaves soft, woolly; faint unpleasant odour. Flowers in few-flowered axillary clusters; calyx green with 10 thin teeth; corolla white, upper lip enclosing stamens (lateral view of flower, a; front view, b).

Common in grass on lower W slope, in shade.

Burma and Thailand through Malesia to NE Australia.

JM 93/63.

Ocimum tenuiflorum L.

(Fl. Mal. 8: 378)

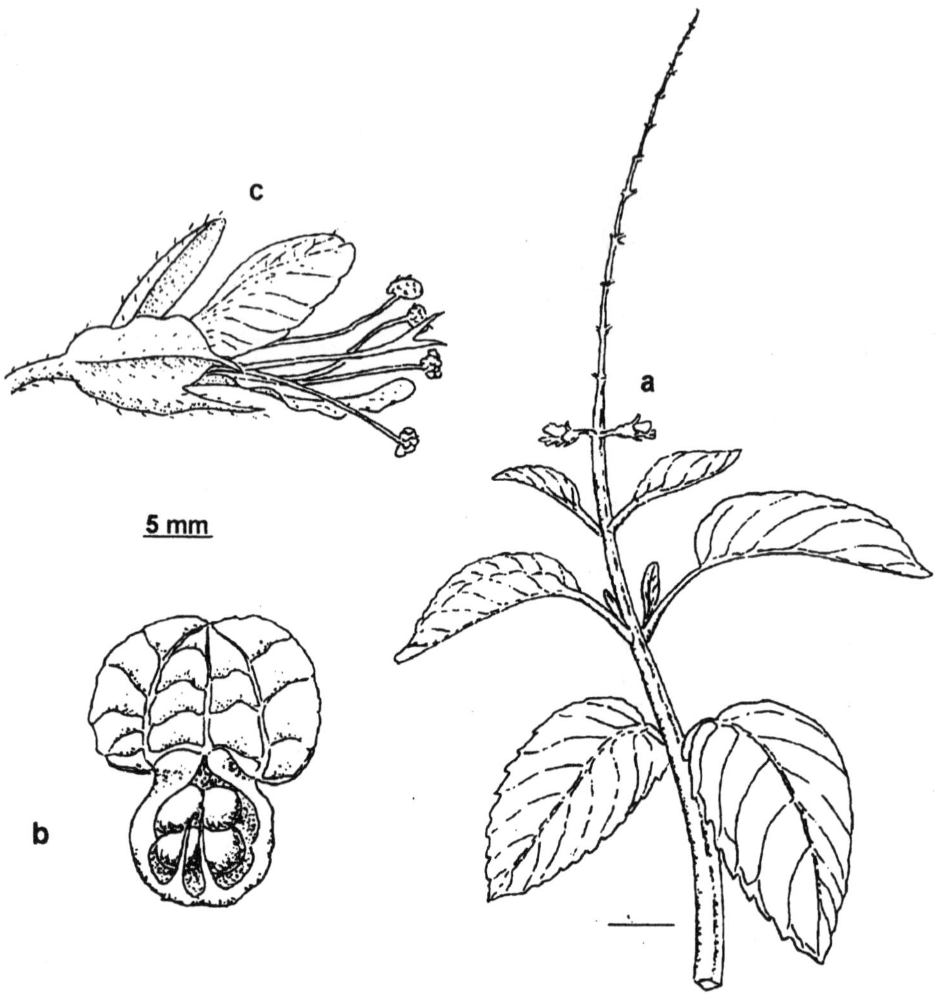

Erect herb, woody at base; mint-like scent. Leaves opposite, petiolate, elliptical or ovate, slightly pubescent; margin shallowly serrate. Flowers in verticillasters in elongate, terminal leafless inflorescence (a); calyx 2-lipped, with a large broad, flat upper lip and a 4-spined lateral-ventral lip (front view, fruiting stage, b); corolla light green, pale blue-violet or white, not enclosing stamens (lateral view of flower, c).

In long grass at N end of island near houses.

Pantropical weed possibly native of tropical Asia.

HCFH 923/13454, 1779/13711; JM 93/56; Kembi K15; Kuduk & Naoni UPNG 6114; Vinas UPNG 4833.

Cassytha filiformis L.

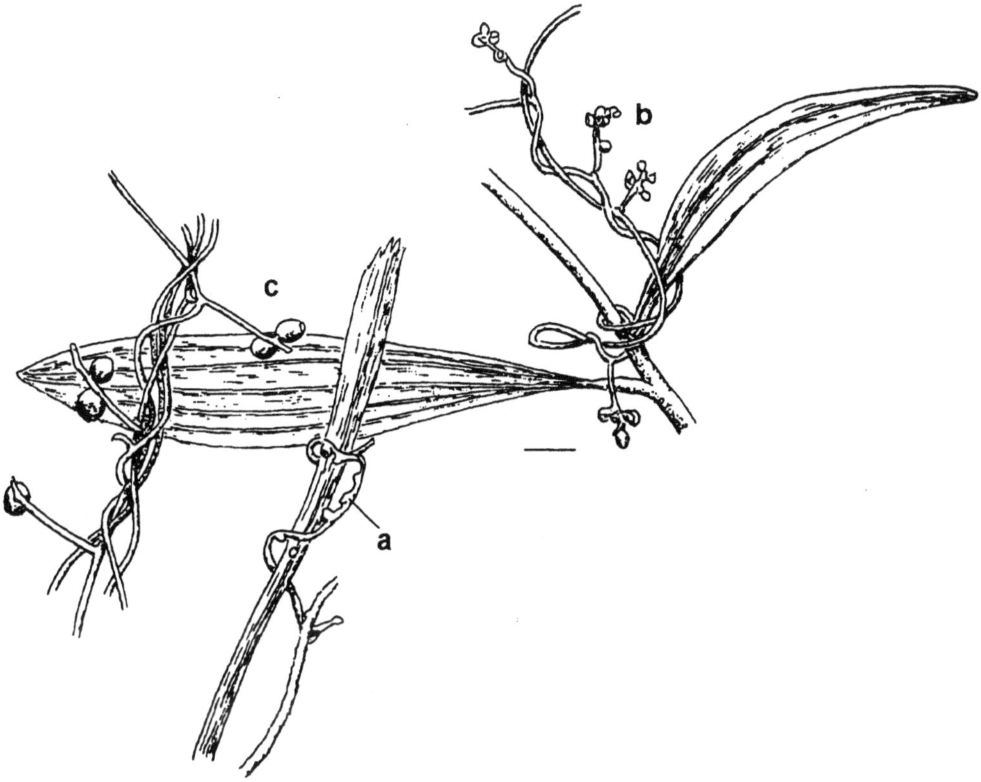

"Leafless", parasitic twining vine shown here climbing over *Acacia auriculiformis*. Stems slender, green or orange; old stems brown, much branched, attached to host plants by small outgrowths (haustoria) (a). Leaves reduced to scales, clasping young stems. Flowers tiny, in short racemes (b), white; corolla 3-lobed. Fruit (c) globose, succulent, greenish-white or orange, translucent, containing a single round seed.

Common in savanna, especially on cycads and grasses, and at scrub edge.

Pantropical.

Frodin UPNG 4291; HCFH 1076/13608; Huxley s.n. 15/6/75; Powell et al. UPNG 2333.

Sometimes put in family Cassythaceae.

Litsea glutinosa (Lour.) C.B.Rob.

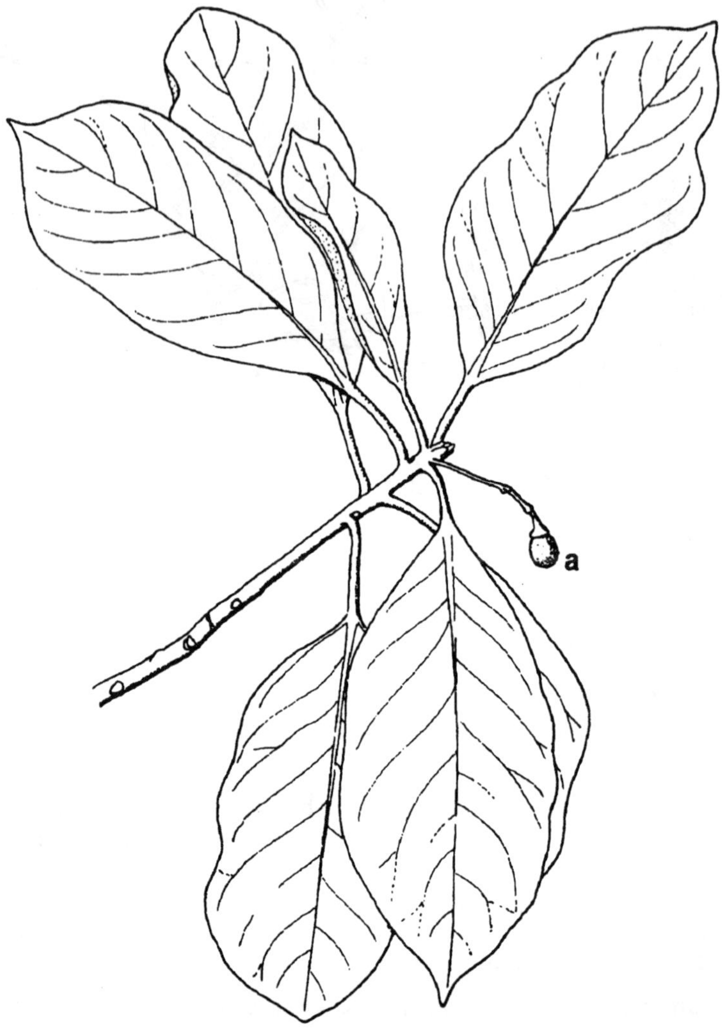

Small tree to 8 m high. Twigs and petioles pubescent, grey, mottled with black. Leaves alternate, dark to mid-green or olive-green above, sometimes tinged with red, yellowish green below, mottled, somewhat leathery; margin entire, slightly undulate. Upper leaf surface ± glabrous; lower surface densely pubescent with creamy brown hairs. Crushed leaves faintly scented. Flowers unisexual, in few-flowered axillary umbels, small, green-cream. Fruit (a) globose, smooth, black, shiny, with white flesh, pleasant fruity smell, and containing a single seed.

Rare at scrub-grassland border, N end and W side of island, to 10 m ASL.

Also in N Australia.

Frodin UPNG 2916; HCFH 1043/13575, 1050/13582; JM s.n. June 91.

Bauhinia binata Blanco

Lysiphyllum binatum (Blanco) de Wit

(Verdcourt: 108, as *Lysiphyllum*)

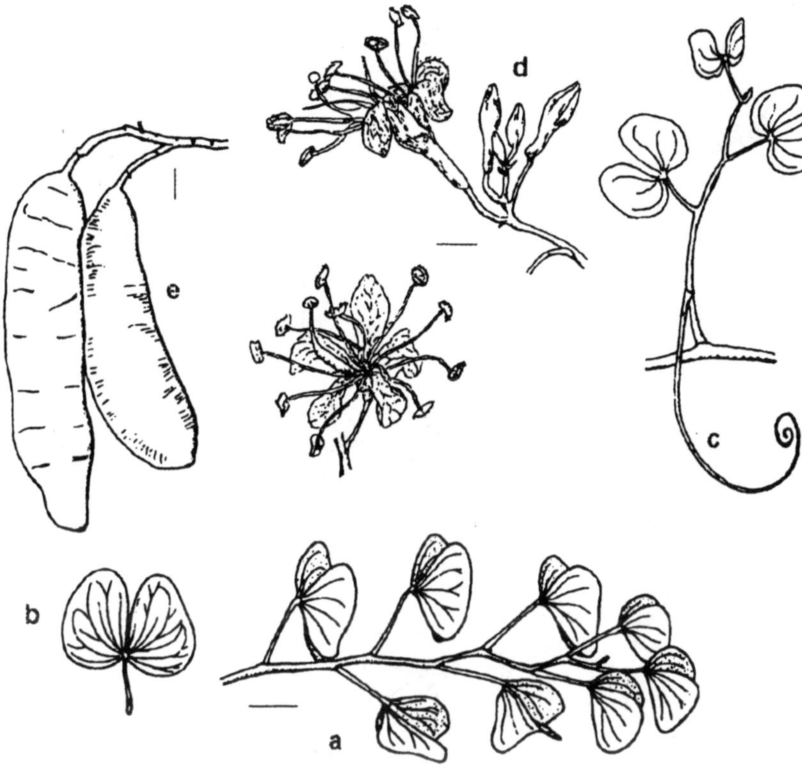

Small spreading tree or scrambling shrub, to 3m high. Twigs slightly zig-zag (a), darkish red-brown. Leaves alternate, 2-partite (b), the blade glabrous, dull, pale glaucous green, angled at apex of yellowish petiole, the two halves closed together much of the time. Flowers in terminal or lateral inflorescences (c); sepals 4-5; petals 5, white; stamens 10, all fertile; anthers reddish. Pods brown or black (d), rattling at maturity, containing several flattened seeds.

Strand, just above high water at N end and E side of island.

Indo-China, Malesia and Australia.

Frodin UPNG 816; HCFH 1729/13661.

Bauhinia variegata L. [not illustrated] is a small tree, 4 m high; leaves rounded, divided for 1/3 of their length into 2 lobes; flowers purple, with 5 stamens, all fertile; fruits narrowly oblong, brown, dehiscent. Native of SE Asia and cultivated in Station area. (See Verdcourt: 119). *HCFH 1792/13744.*

Caesalpinia bonduc (L.) Roxb.

(Verdcourt: 23)

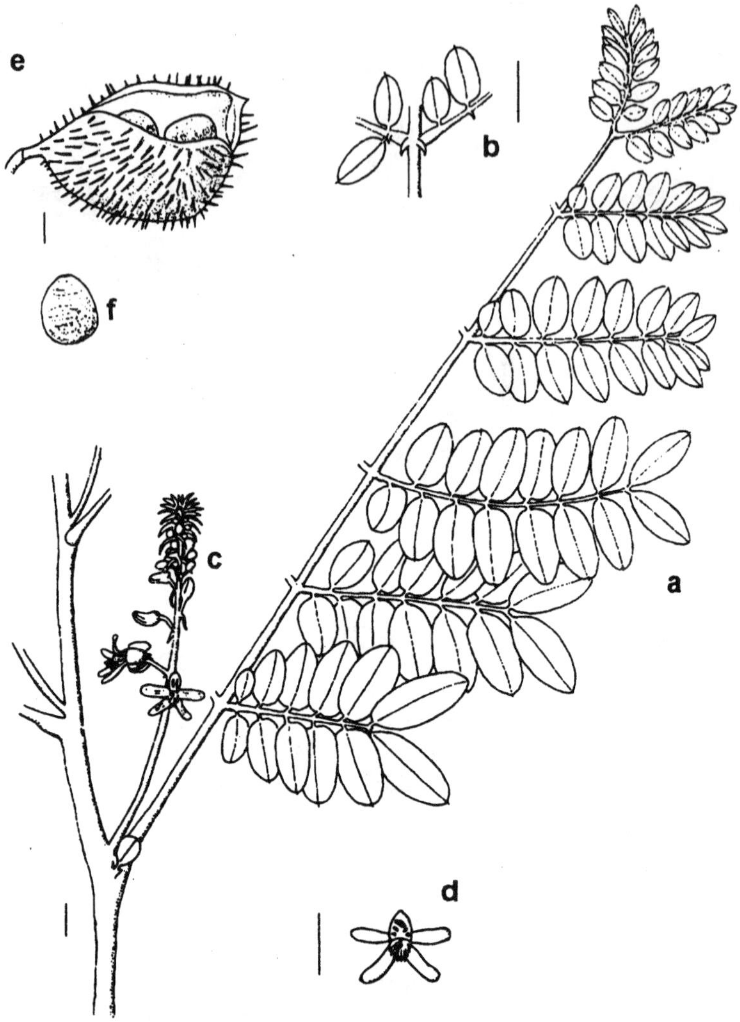

Climber or scrambling shrub. Leaves bipinnate (a), alternate, pinnae 3-9+ pairs, leaflets 6-9+ pairs per pinna; main rachis puberulous; leaflets glabrous, mucronulate. Paired spines on main rachis at junction of pinnae (b). Flowers in a raceme (c), inserted just above leaf axil; buds brown, velutinous. Petals 5, yellow, with red markings on standard (d). Pods green turning brown, densely spiny on outside (e), dehiscent, containing 1-3 globose seeds, the testa very hard, grey, with cracked-glass surface (f).

Strand and edge of coastal scrub on E & W sides, and N end.

Widespread in tropics.

HCFH 610/13141, 1038/13570.

Chamaecrista absus (L.) Irwin & Barneby

Cassia absus L.

(Verdcourt: 36; Henty & Pritchard: 116, as *Cassia*)

Erect, branched annual herb to 50 cm tall, slightly sticky due to minute glandular hairs. Leaves paripinnate; leaflets markedly asymmetric, in two pairs only. Inflorescence a sparse terminal raceme. Flowers (a) with five pale bright pink petals, the standard slightly larger than the others. Stamens 5, very unequal. Fruit (b) flattened, about 5 cm long.

Uncommon, seen only on the ridge top, among tall grass.

Widespread in the Old World tropics. In New Guinea, mainly in dryer regions.

JM 95/145.

Chamaecrista nictitans (L.) Moench

Cassia lechenaultiana DC.

C. mimosoides L. subsp. *lechenaultiana* (DC.) Ohashi

(Verdcourt: 48, as *Cassia*)

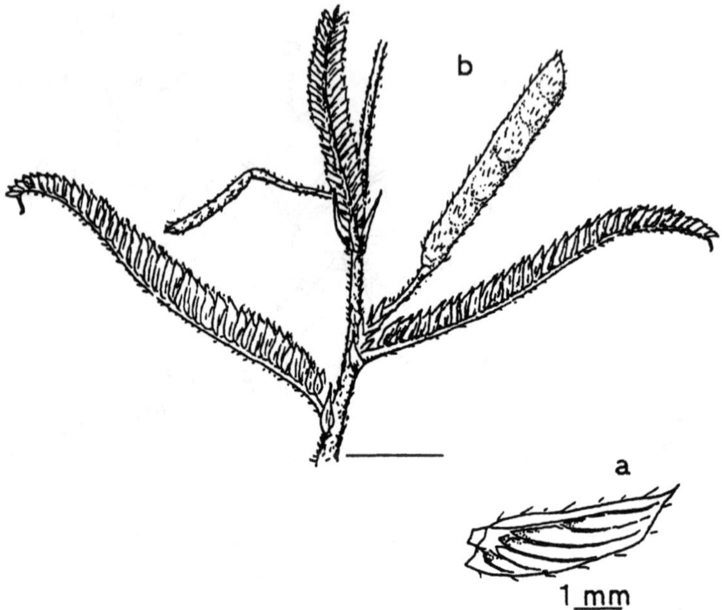

Erect, annual, woody herb to 0.5 m, hirsute all over. Leaves pinnate, terminating in a bristle, not a leaflet; leaflets (a) < 40 pairs, 4-5 mm long. Flowers solitary to few in the axils; calyx 2-lipped; petals yellow, 9 mm long. Fruits flattened, about 12-seeded, c. 30 mm long (b).

Uncommon, in grass near ridge top.

India to SE Asia, Malesia and some Pacific islands.

JM 93/73.

Intsia bijuga (Colebr.) Kuntze [not illustrated] was recorded by Frodin as "only once seen" on the shoreline behind mangroves: tree with pinnate leaves; leaflets 1 or 2 pairs, c. 10 x 4 cm, obliquely ovate, slightly falcate. Widespread, E Africa to Malesia. (See Verdcourt: 91). *Frodin UPNG 3187.*

Senna siamea (Lam.) Irwin & Barneby [not illustrated] (syn. *Cassia siamea* Lam.), is a tree 4 m high, cultivated near the Wet Lab.: leaves pinnate; leaflets 4-16 pairs, elliptical, 3-7.5 x 1.2-1.6 cm, on short petiolules; inflorescence erect; flowers numerous; petals yellow; fruits 15-30 x 1.2-1.6 cm, dehiscent. Native of SE Asia. (See Verdcourt: 52). *HCFH 1793/13745.*

Peltophorum pterocarpum (DC.) Heyne

(Verdcourt: 16)

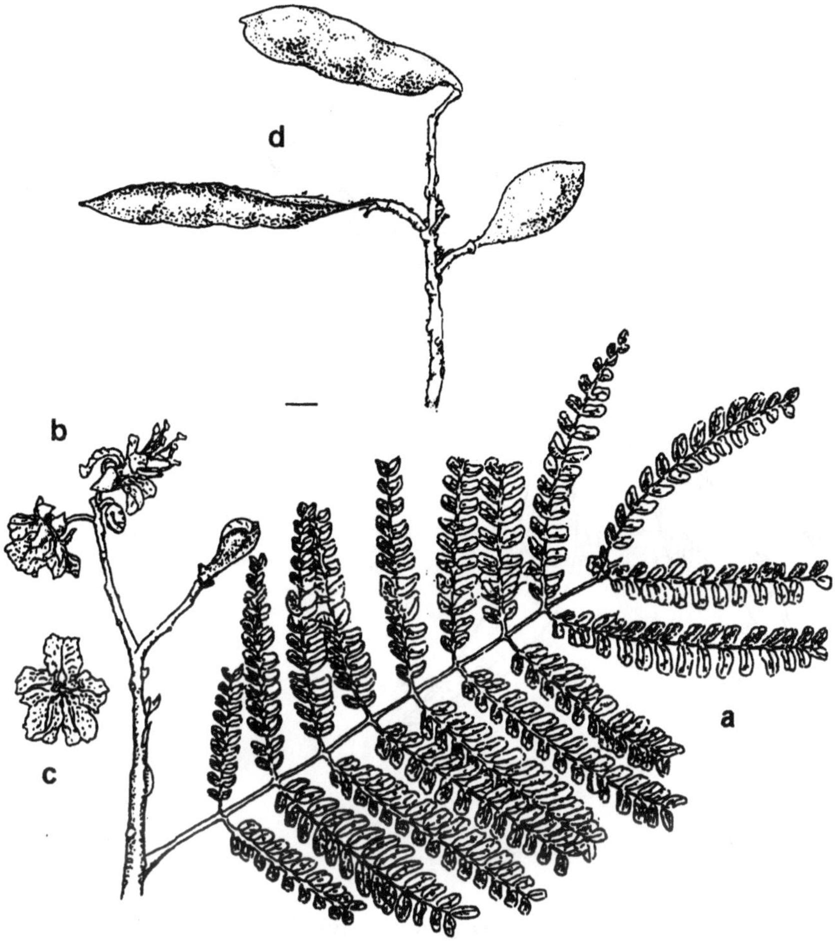

Tree to 8+ m tall. Leaves (a) alternate, bipinnate; pinnae 7-15 pairs; leaflets 8-20 pairs, mid-green, rather dull. Young stems, petioles and inflorescence stalks covered in rusty brown pubescence. Inflorescence an erect terminal raceme (b). Sepals yellow-green; petals yellow with central brown rib, much crinkled (c); gynoecium green. Fruits red-brown (d), woody, usually one-seeded, with longitudinal striations, eventually splitting down middle of valve.

Mangrove, NE corner of island, and cultivated in Station area.

Native of tropical Asia and Australia.

Frodin UPNG 5207; HCFH 1796/13748.

Acacia auriculiformis A. Cunn. ex Benth.

(Fl. Mal. **11**: 58; Verdcourt: 163)

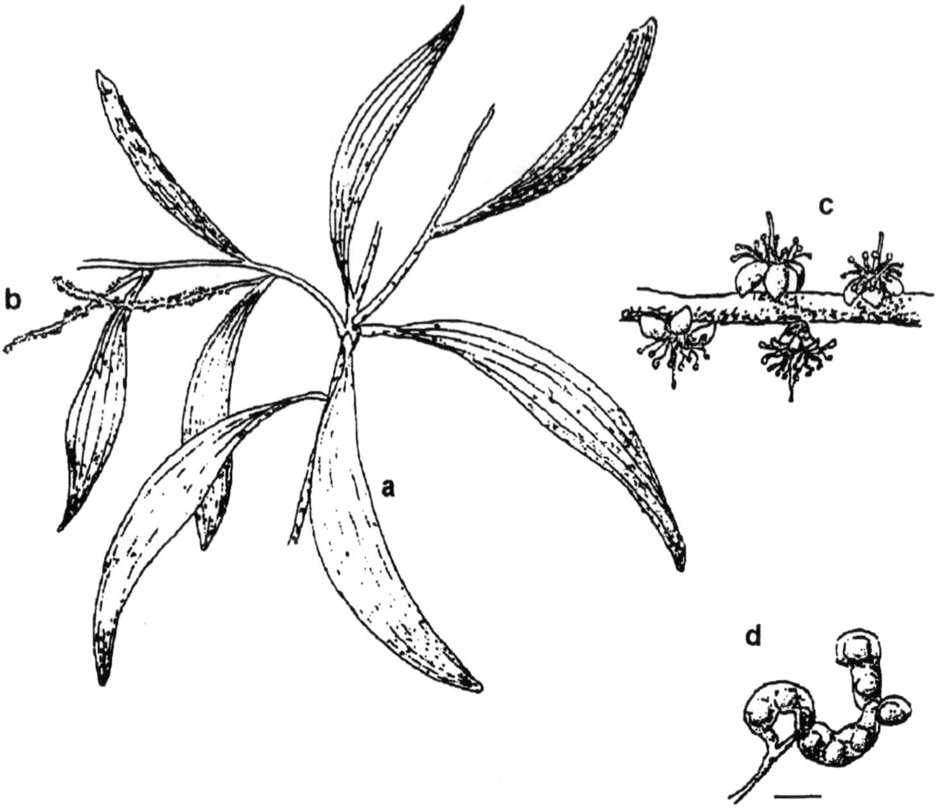

Small tree to 6 m tall. Stems green to brown, with many lenticels. Leaf-like phyllodes (a) with several parallel longitudinal veins, dark olive-green, slightly glossy, bearing a gland near base. Inflorescence an axillary spike (b); flowers very small, bright yellow, with numerous stamens (c). Fruits red-brown, rather woody, flat, curled, dehiscent (d).

Savanna, especially W side and N end of island.

New Guinea and N Australia.

HCFH 1087/13619; Kuduk & Naoni UPNG 6122.

Albizia carrii Kanis

(Fl. Mal. **11**: 71; Verdcourt: 181)

Semi-deciduous tree 10-15+ m high. Twigs brown. Leaves bipinnate (a); pinnae 2-4 pairs; leaflets 5-10 pairs, darkish or mid-green above, paler below, slightly pubescent; apex minutely mucronulate. Petiole dull green, with circular gland. Inflorescence an axillary umbel (b). Flowers hermaphrodite or male. Central flower on a short pedicel, surrounded by ring of 10 on longer pedicels. Corolla 5-lobed, greenish white; stamens numerous, with a pinkish caste. Pods oblong, c. 20-26 x 3.6-4.5 cm, with raised transverse veins, puberulous when young, pale brown (c).

Monsoonal woodland on hill side.

Endemic to Port Moresby area.

Frodin UPNG 854, 855; HCFH 1710/13642, 1797/13749.

Leucaena leucocephala (Lam.) de Wit

L. glauca sensu Benth.

(Fl. Mal. **11**: 182; Verdcourt: 154)

Shrub. Older stems with numerous lenticels (a). Leaves alternate, bipinnate; pinnae 3-8 pairs, leaflets 7-17 pairs, asymmetrical, obliquely oblong-lanceolate (b), mid-green above; prominent gland on rachis between basal pair of pinnae. Inflorescence a pale yellow ball, c. 1.3 cm diameter (c). Corolla greenish-white; stamens numerous, yellow (d). Pods with papery valves, glabrous or puberulous, splitting down both sutures (e). Seeds brown, flat, c. 20 per pod.

Edge of savanna, near Caretaker's house.

Native to America; widespread in tropics, on roadsides, wasteground and plantations.

B. Lawong s.n. 24/8/88.

The following cultivated species, which are not illustrated, grow in the Station area:

Adenanthera pavonina L. Small tree; leaves bipinnate with 3-5 pairs of pinnae; leaflets 5-9 down each side of secondary rachises, elliptical, 1.5-4.5 x 1.2-2.3 cm; flowers tiny, in a long spike, creamish; pods dehiscent, curled; seeds circular, bright red. Widespread in tropical Asia and Malesia. (See Fl. Mal. **11**: 173; Verdcourt: 138).

Pithecellobium dulce (Roxb.) Benth. Tree, much branched; twigs slightly zig-zag, bearing a pair of stipular thorns at each node; leaves bipinnate with 1 pair of pinnae, each pinna composed of 2 asymmetrically elliptical leaflets, 0.7-5 x 0.3-2.3 cm. Flowers tiny, in cream-coloured heads c. 1 cm diameter. Fruits spirally twisted, with pulpy aril around seeds. Native of S America; widely cultivated in tropics. (See Fl. Mal. **11**: 155; Verdcourt: 209). *Bigilale BH18; Frodin UPNG 3199.*

Samanea saman (Jacq.) Merr. Rain Tree. Medium-sized tree with spreading crown; leaves bipinnate; pinnae 2-6 pairs, leaflets 3-8 pairs, asymmetrical and somewhat rhomboid-elliptical, the rachis bearing a gland between each pair of leaflets. Flowers in pedunculate semi-hemispherical clusters, pink. Fruits brown or black, narrowly oblong, 10-20 x 1.6-2 cm, containing a sweet pulp. Cultivated, near Old Hostel. Native of C and S America; widely cultivated in tropics. (See Fl. Mal. **11**: 156; Verdcourt: 207).

Abrus precatorius L.

(Verdcourt: 305)

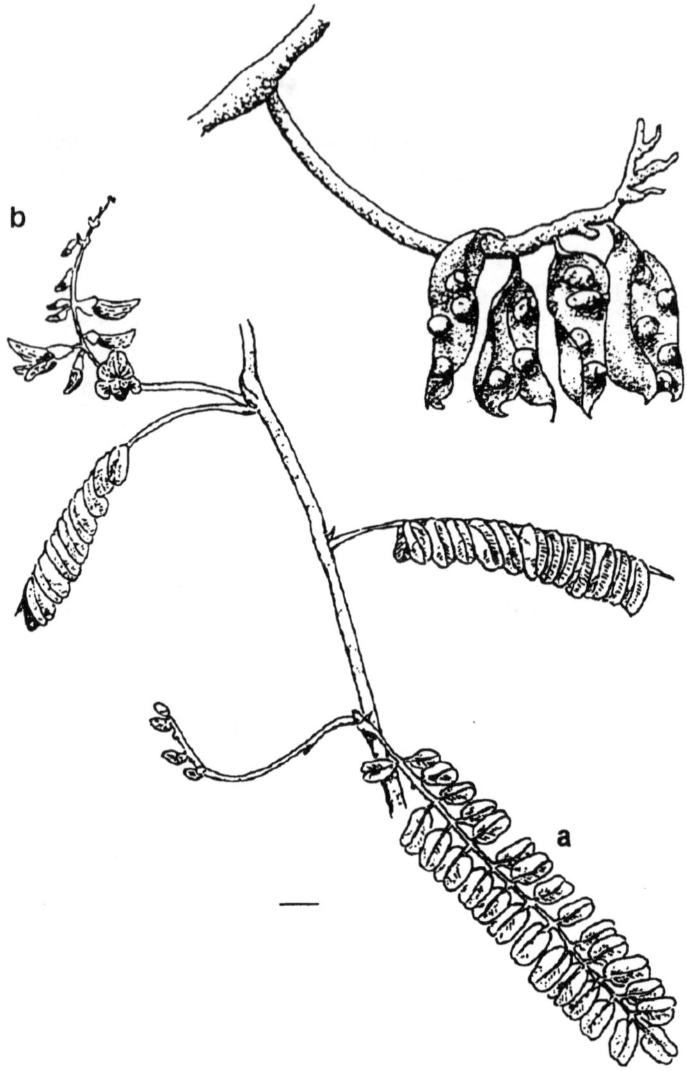

Woody climber. Leaves pinnate (a); leaflets 12-14 (-34) pairs, ± glabrous, mid-green above, paler below. Inflorescence an axillary raceme (b). Flowers deep blue or purple. Pods oblong, dehiscent, brown, containing several shiny seeds (c), each red with black base.

Monsoonal scrub, near base of nature trail, E side and elsewhere.

Widespread in tropical Asia, Australia and the Pacific; another subspecies in Africa. Introduced elsewhere in tropics.

Frodin UPNG 845, 4244, 5224; HCFH 1735/13667; Powell et al. UPNG 2348.

Alysicarpus bupleurifolius (L.) DC.
Alysicarpus vaginalis (L.) DC.

(Verdcourt: 420)

Decumbent herbs with ascending shoots. Leaflets one, variable in shape from oblong-elliptical (a), narrowly elliptical (b) to linear-lanceolate (c), 0.5-8 x 0.3-2.6 cm, glabrous to puberulous. Stipules papery, persistent, ± sheathing stem. Inflorescence (d) long, lax, terminal or leaf-opposed; flowers paired; calyx papery; standard purple to orange or yellow; wings variable, often mauve. Pods linear-oblong, slightly constricted between articles (*A. bupleurifolius*) as illustrated above, or not constricted (*A. vaginalis*).

Station area, 1 m ASL and on ridge top. In savanna and grassland.

Asia and Malesia (*A. bupleurifolius*) or widespread throughout tropics (*A. vaginalis*).

Frodin et al. UPNG 5210, 6671 (*A. bupleurifolius*).
Frodin UPNG 4275, 5231, 6672; HCFH 1782/13714 (*A. vaginalis*).
JM 94/106 (sp. not determined).

Atylosia scarabaeoides (L.) Benth.

A. reticulata auct. non (Aiton) Benth.

(Verdcourt: 540)

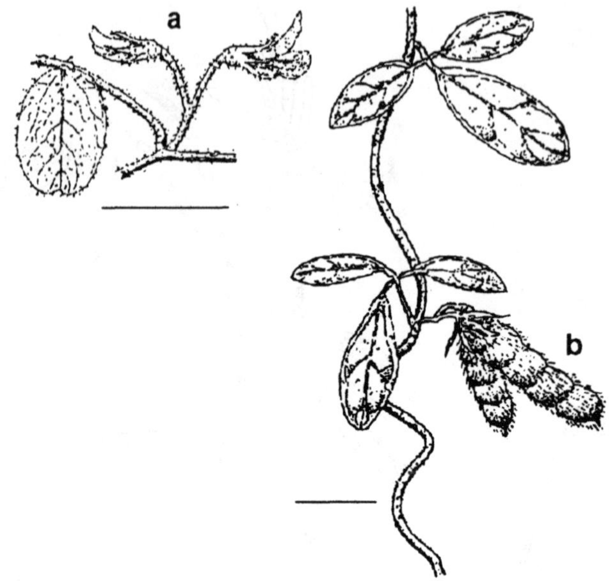

Climber or trailer. Stems often reddish. Leaves trifoliate, reported as having resinous gland dots on underside, but obscured by dense, grey, velutinous pubescence; leaflets ± elliptical, terminal one ± rhomboidal. Flowers in short, penduncülate, axillary clusters (a); standard yellow, flushed with crimson; wings orange; keel green. Pods oblong, hairy, slightly inflated, indented between each seed (b).

Frequent in grassland and scrub edge, E side of island, and in savanna on ridge.

Widespread in Asia and Australia; introduced in Africa.

Frodin UPNG 4288, 5252; JM 93/72.

Canavalia papuana Merr. & Perry
(Verdcourt: 474)

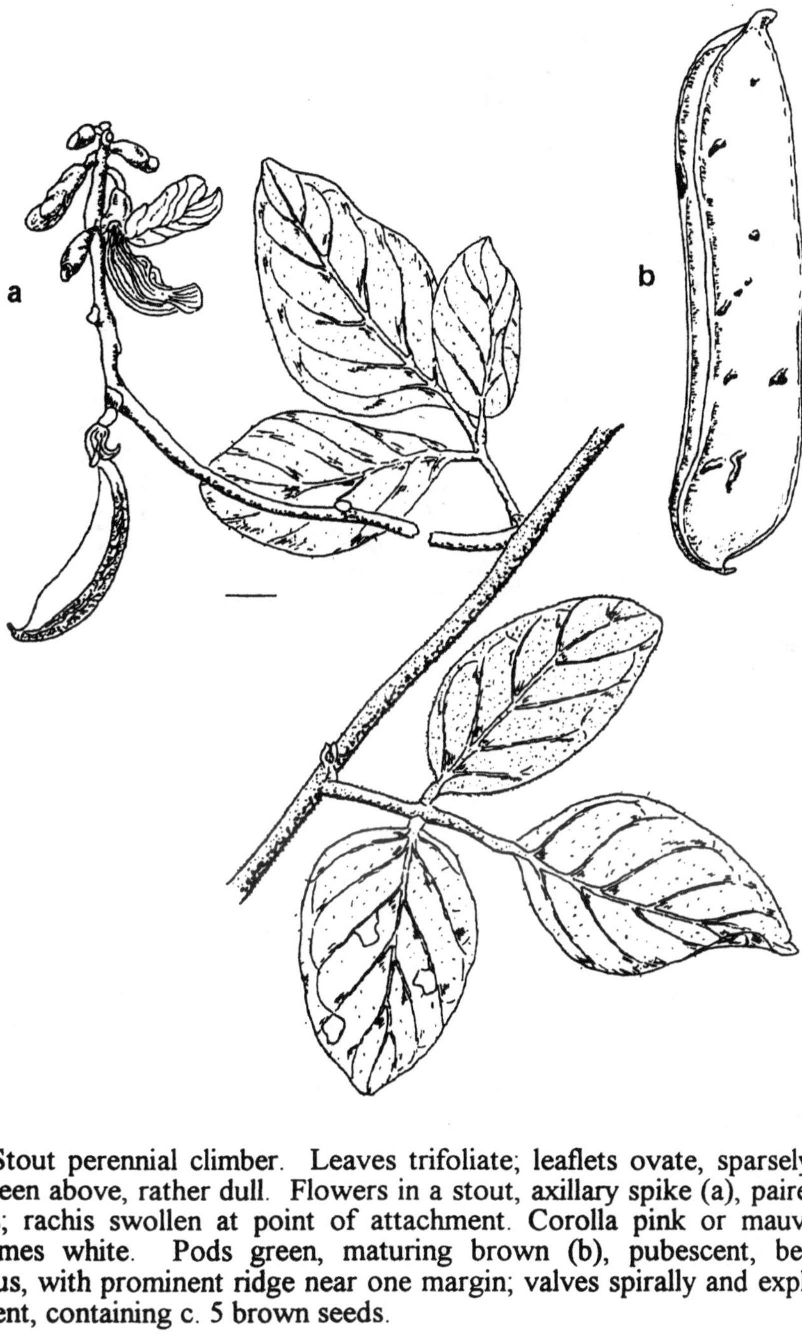

Stout perennial climber. Leaves trifoliate; leaflets ovate, sparsely hairy, mid-green above, rather dull. Flowers in a stout, axillary spike (a), paired or in groups; rachis swollen at point of attachment. Corolla pink or mauve, keel sometimes white. Pods green, maturing brown (b), pubescent, becoming glabrous, with prominent ridge near one margin; valves spirally and explosively dehiscent, containing c. 5 brown seeds.

Twining over grasses and shrubs in savanna on hillside.

New Guinea and N Australia.

Frodin UPNG 3168, 3169, 4276, 5266; HCFH 1730/13662; Huxley s.n. 15/6/75; Powell et al. UPNG 2329.

Christia obcordata (Poir.) Bakh. f. in Steenis

(Verdcourt: 419)

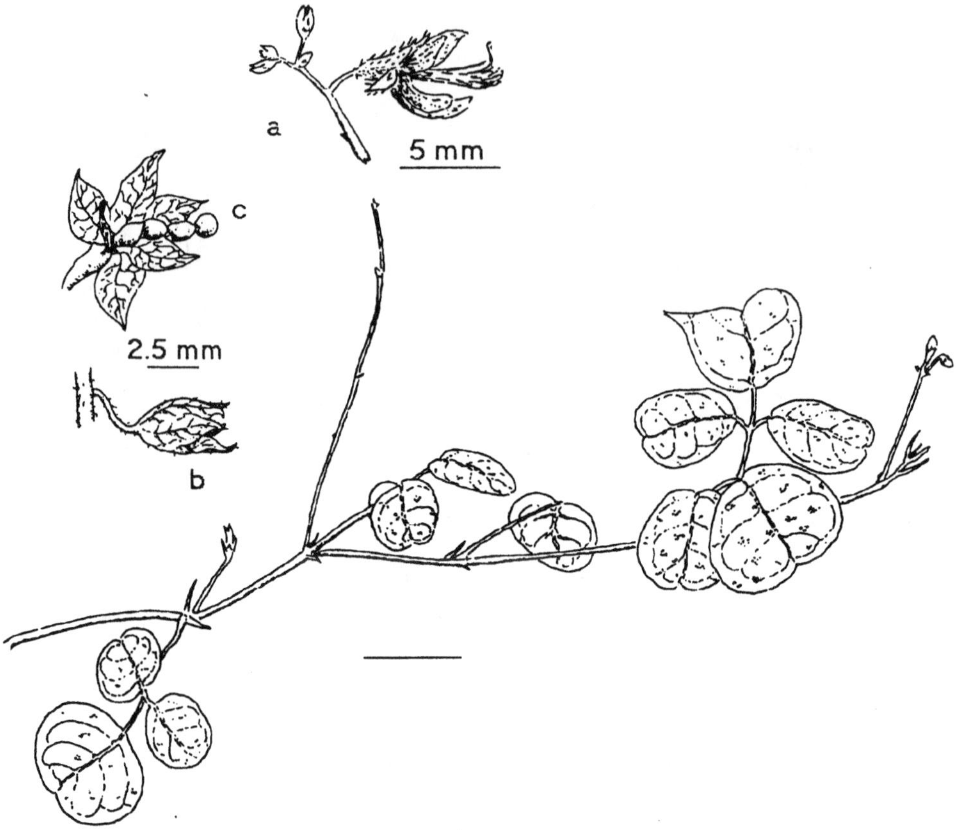

Prostrate or creeping herb. Leaves trifoliate or lower ones unifoliate. Terminal leaflet wider than long, i.e. transversely elliptical or obreniform, emarginate at apex; lateral leaflets rounded or oblong, emarginate; leaflets mid-green above, paler and sparsely pubescent below. Flowers in lax terminal or axillary racemes (a); calyx (b) papery and veined, enlarging in fruit; corolla purple, occasionally white. Pods (c) with 2-5 joints, constricted between seeds, folded inside calyx while developing, expanding later, glabrous or hairy.

Savanna on hillside and Station area.

E and SE Asia, Malesia.

Frodin et al. UPNG 4283, 6659; JM 94/111.

Clitoria ternatea L.

(Verdcourt: 502; Henty & Pritchard: 120)

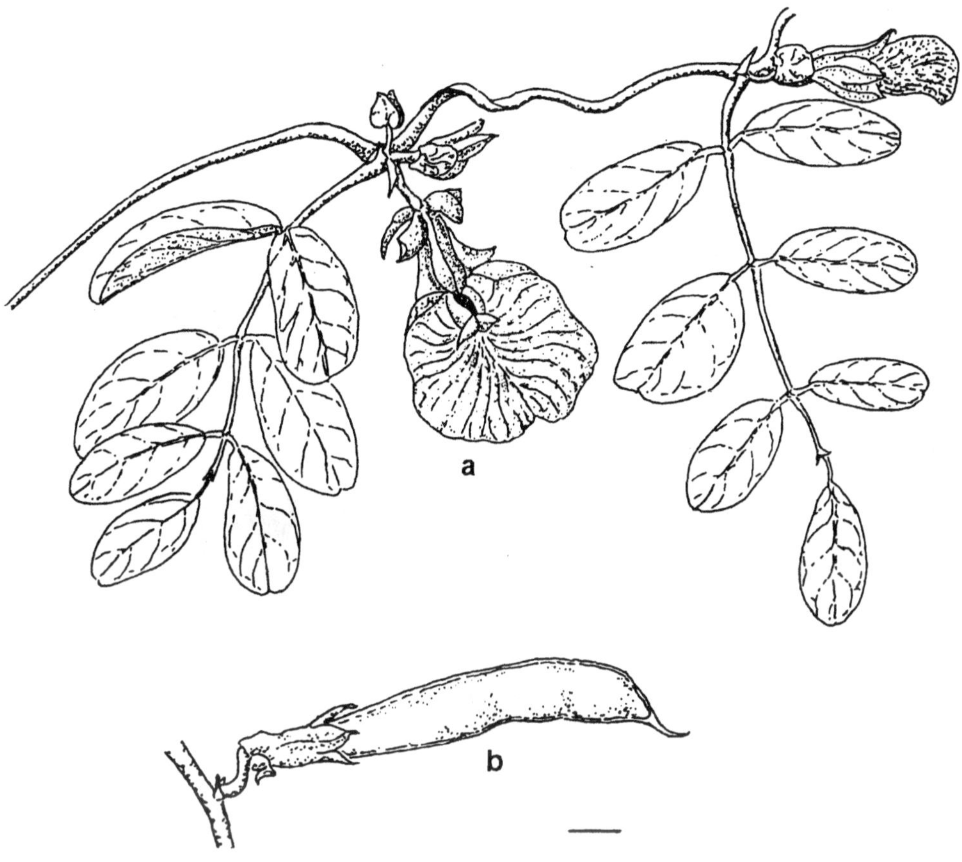

Climbing or scrambling herb. Leaves pinnate; leaflets 5-7, ovate or elliptical, sparsely hairy, pale to mid-green above. Flowers axillary (a), 2-3 per axil but opening singly; corolla showy, standard held lowermost, blue or pale blue. Fruit fawn-brown (b), spirally dehiscent; valves pubescent; seeds black.

N end near generator, in long grass.

Pantropical weed, original distribution obscure.

HCFH 1795/13747.

Crotalaria montana Roth

(Verdcourt: 581)

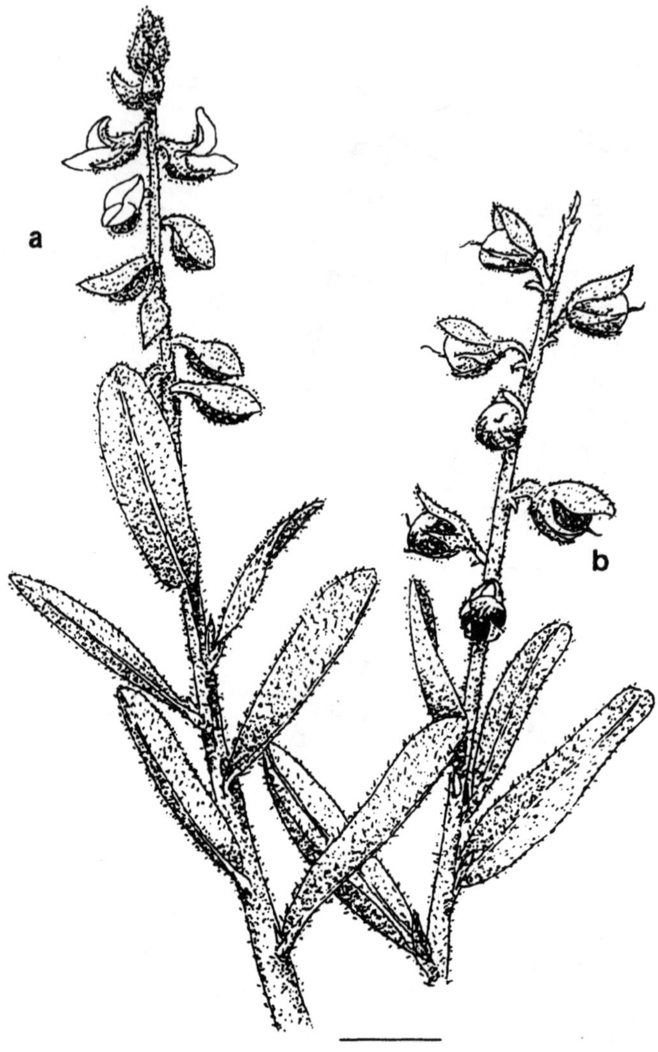

Erect branched herb. Whole plant densely covered in whitish or pale brown hairs, so plant has grey or red-brown cast. Leaves simple, oblong-elliptical, 1.3-5 x 0.3-1 cm. Flowers in terminal racemes (a). Calyx densely hairy; corolla yellow. Pods tiny, inflated, black (b).

Widespread in grass.

Asia, Malesia to N Australia.

Frodin UPNG 4278, 5256; HCFH 1780/13712.

Desmodium umbellatum (L.) DC.

(Verdcourt: 410)

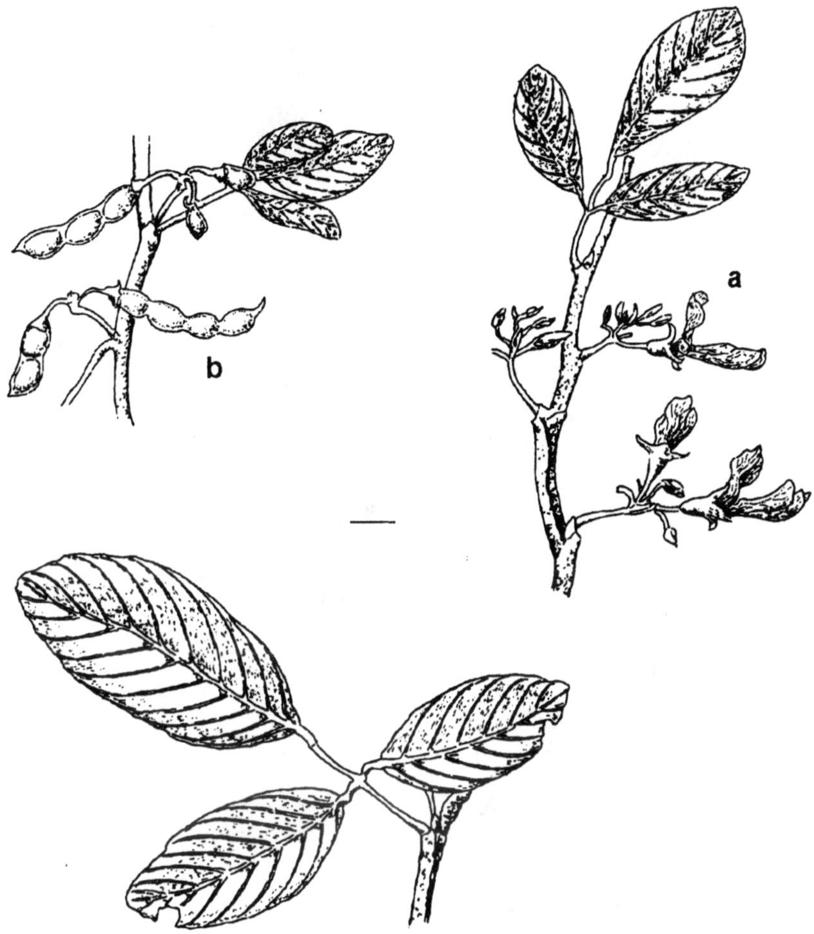

Shrub or small tree to 4 m high. Stems squarish, green-brown. Leaves trifoliate; leaflets elliptical or obovate, glabrous above and minutely hairy below; margins slightly sinuate and reddish. Flowers in short axillary umbel-like inflorescences (a); corolla white. Fruits red-green, almost woody when mature, 2-7 articles, pubescent (b).

Common on ridge and W slope at S end of island.

E Africa, Asia, Malesia, to N Australia.

Frodin UPNG 5203; HCFH 1079/13611.

Desmodium velutinum (Willd.) DC.

(Verdcourt: 411)

Woody herb to 40+ cm high. Leaves unifoliate or, if trifoliate, then terminal leaflet much larger than lateral ones; leaflets ovate or round, sparsely and roughly hairy above, velutinous below; margin slightly sinuate. Flowers in dense terminal spikes (a), each flower subtended by a bract densely covered with hooked hairs (b); corolla pinkish-mauve to blue. Fruits of 2-7 articles, densely covered in hooked hairs.

Occasional in savanna on hillside and along nature trail.

Widespread from E Africa, SE Asia to Malesia.

HCFH 627/13158, 1794/13746.

Erythrina variegata L.

(Verdcourt: 425)

Deciduous tree to 8+ m high. Bark with longitudinal stripes and scattered prickles. Leaves trifoliate, terminal leaflet broader than long, ovate to ± rhomboidal, glabrescent, pale yellow-green and semi-glossy above, greyer below. Flowers in a spike 30-40 cm long (a) with scurfy indumentum. Corolla bright red (b); nectar abundant. Pods elongate, sausage-shaped, somewhat constricted between seeds (c); valves veined, with scurfy indumentum; seeds bean-shaped, pale wine-red (d).

Occasional in mangrove and strand, NE and E side of island.

From E Africa, S Asia, Malesia to Polynesia.

HCFH 1083/13615.

Flemingia strobilifera (L.) R.Br. [not illustrated]

(Verdcourt: 550)

Shrub 1.5 m high. Stems pubescent. Leaves unifoliate, ovate, ± glabrous above, velutinous below; light olive-green above, paler below; margin sinuate. Flowers in axillary inflorescence with large overlapping leafy-papery bracts; rachis of inflorescence zig-zag. Corolla cream-green. Fruits ellipsoid.

At edge of grassland on hill behind houses; collected in 1975-76; not seen in 1990.

Asia, Malesia; widely grown in tropics and naturalized in Pacific.

Frodin UPNG 3177; Huxley s.n. 15/6/75.

Erythrina variegata L.

Galactia tenuiflora (Willd.) Wight & Arn.

(Verdcourt: 478)

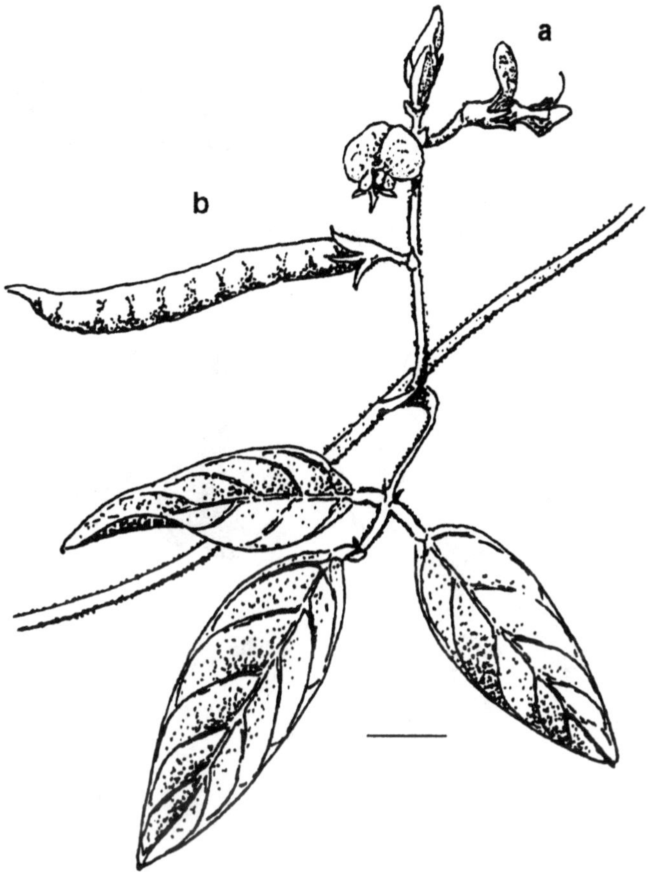

Slender climbing or trailing herb. Stems pubescent. Leaves trifoliate; leaflets ovate or elliptical, 2-6.5 x 0.3-3 cm, dull green and velutinous above, paler and pubescent below. Inflorescence axillary, lax, c. 10 cm long, the groups of flowers equidistant along its length. Flowers small (a); upper 2 calyx lobes completely fused; corolla delicate, purplish or pinkish, drying blue. Pods linear-oblong (b), c. 5 x 0.5 cm, spirally dehiscent; valves finely pubescent; seeds black.

Savanna, by nature trail and on ridge top.

Africa, Asia, Malesia.

Frodin et al. UPNG 2934, 4277, 5254; HCFH 2000/13752.

Glycine tomentella Hayata

(Verdcourt: 492)

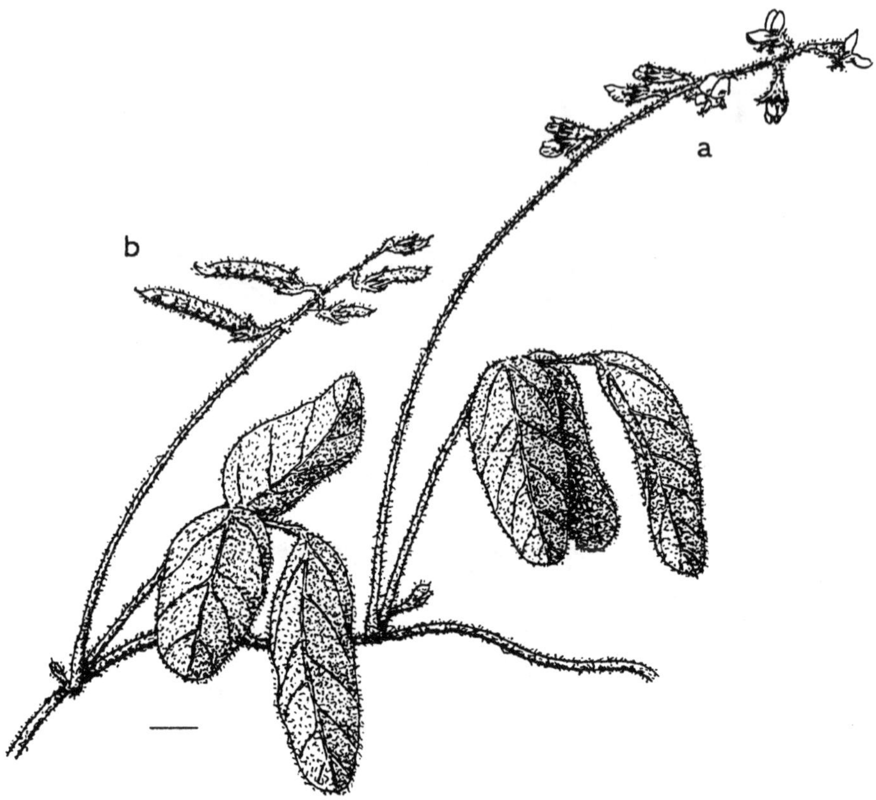

Twining or trailing herb. Stems covered with long ferruginous hairs. Leaves trifoliate; leaflets elliptic-oblong, c. 1.5-5 x 0.7-2.7 cm, sparsely to velvety pubescent, especially below, often with a greyish cast. Inflorescence axillary, the flowers clustered towards the top (a); upper 2 calyx lobes only partly fused; corolla pinkish-purple, standard paler, blue when old, drying orange. Pods linear-oblong, up to 2.5 cm long, dehiscent; the valves pubescent (b).

Station area and savanna, by nature trail and ridge top.

Also in Australia, Philippines and Taiwan.

Frodin et al. UPNG 5208, 6666, s.n. 1972; HCFH 1799/13751; Tippett UPNG 826.

Rather similar to the preceding species, and often found growing together. However, in *Galactia tenuiflora* the upper 2 calyx lobes are completely fused, i.e. it has only 4 calyx teeth, while *Glycine* has 5 (D. Frodin, pers. comm.).

Indigofera colutea (Burm. f.) Merr.

I. viscosa Lam

(Verdcourt: 351)

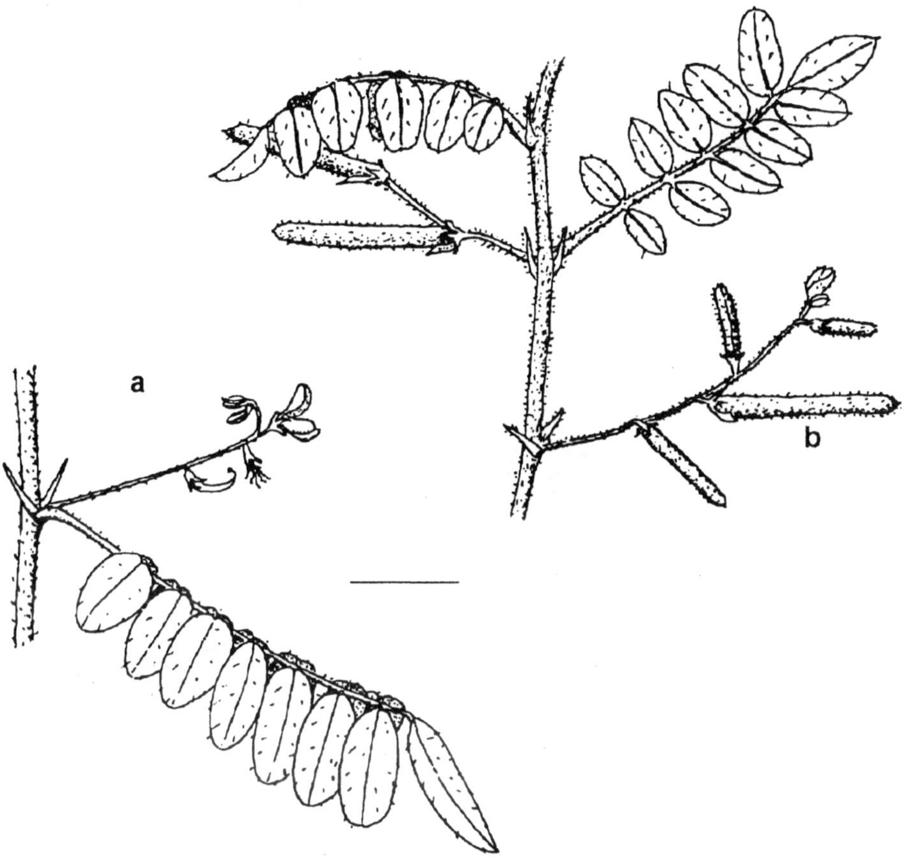

Erect, annual, woody herb to 0.5 m. Stems and fruits with forked hairs and simple glandular hairs. Leaves imparipinnate; leaflets 7-15, up to 1.4 x 0.5 cm. Flowers minute in elongate, few-flowered, axillary racemes (a). Corolla red and green. Fruit cylindrical (b), only slightly flattened, < 2 cm long.

Among grass and other ground flora, in shade, E side.

Africa to India and Australia.

JM 93/71, 94/103.

Indigofera hirsuta L.

(Verdcourt: 351)

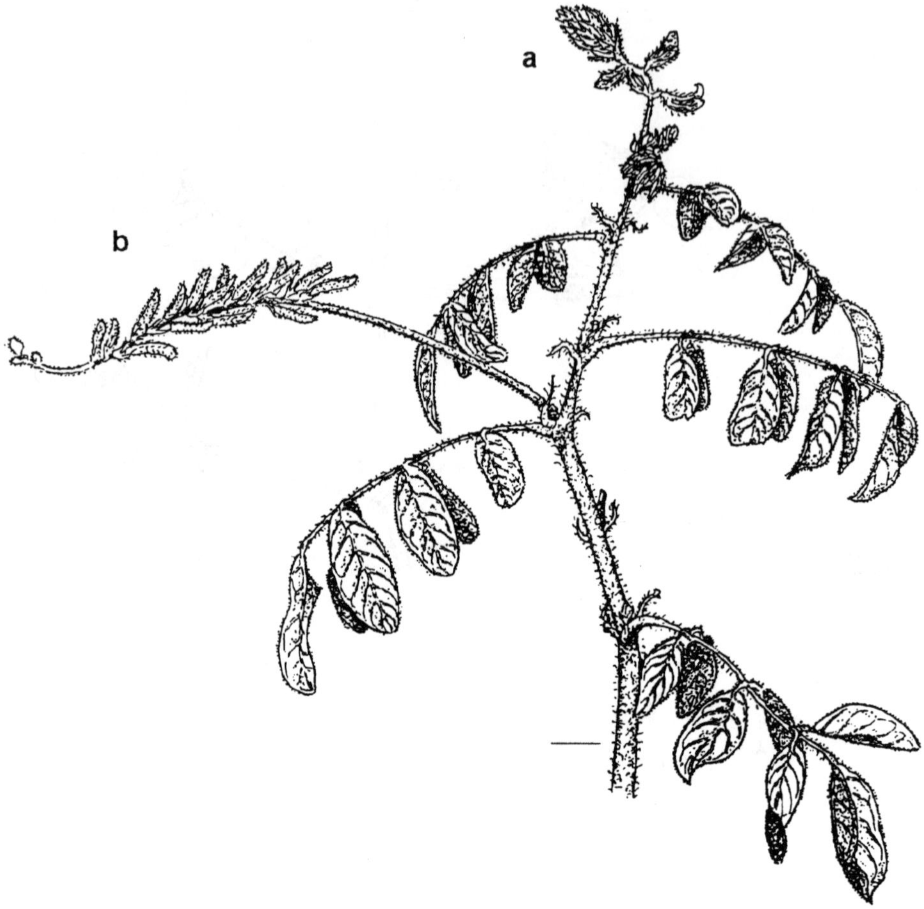

Branched, erect, woody herb to 1.5 m, covered with white hairs. Stems and petioles reddish. Leaves imparipinnate; leaflets to 3 x 1 cm. Flowers very small (a) in dense terminal or axillary racemes. Corolla reddish. Fruits 1.5 cm long, hairy (b).

On W slope and ridge top.

Africa, Madagascar and S Asia to N Australia.

JM 93/72.

Indigofera linifolia (L.f.) Retz.
(Verdcourt: 353)

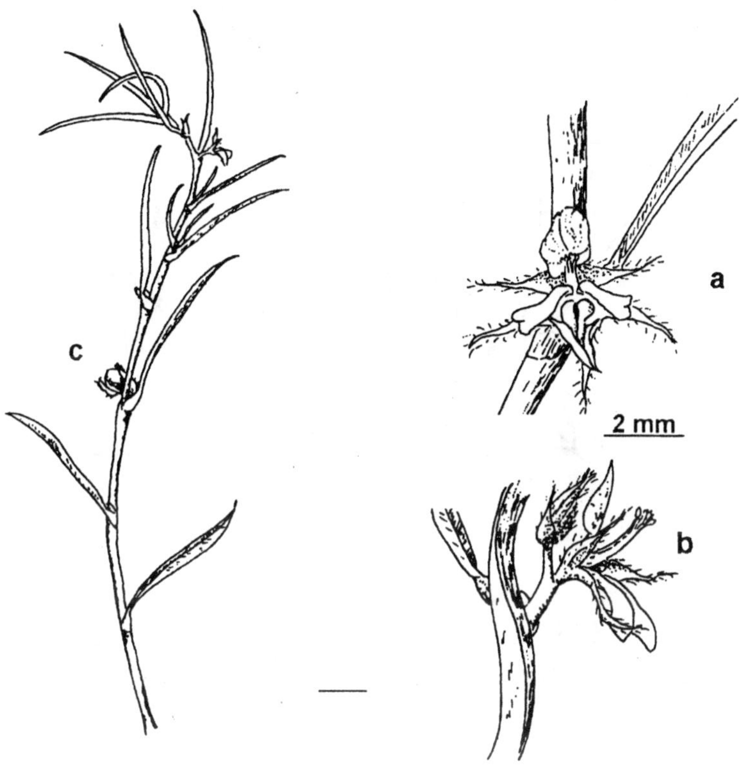

Branched erect herb to c. 40 cm, covered with dense white pubescence. Leaves simple, linear, to c. 40 x 2 mm, greyish-white; hairs on upper surface oblique and on lower surface parallel to margins. Flowers in few-flowered, almost sessile axillary clusters, small (a, front view; b, lateral view); corolla reddish. Fruit (c) globose, tiny, covered in dense white pubescence.

Occasional in savanna on ridge.

E Africa, Asia, Malesia to Australia.

Frodin UPNG 4289, s.n. March 1972.

Indigofera linnaei Ali

(Verdcourt: 353)

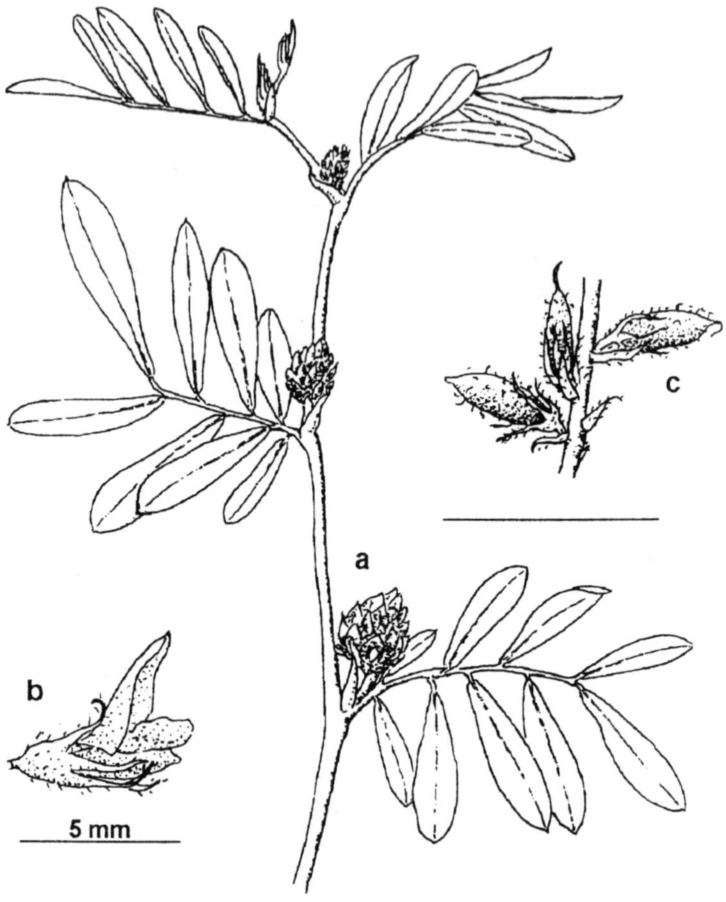

Ascending herb to c. 30 cm, covered with dense white pubescence. Leaves pinnate; leaflets c. 10, alternate, oblanceolate, to 1.2 x 0.4 cm, with white pubescence on both surfaces. Inflorescence a dense, hairy, axillary cluster, c. 1.5 cm long (a); flower, lateral view (b); corolla red. Fruit oblong-cylindrical, to 4.5 mm long, pubescent (c).

Occasional at edge of long grass around Station area and in savanna on ridge top.

India to Australia.

Frodin et al. UPNG 4279, 6668, 6669, s.n. March 1972; HCFH 1781/13713; Lawong s.n. 24/8/88.

Plate 1. Motupore Island, east aspect. The shore bears a fringing mangrove, mostly *Rhizophora* spp. *Eucalyptus* savanna on the slopes and crest. Dry season.

Plate 2. Motupore Island, west aspect. The rocky shore has a sparse mangrove. Monsoon scrub with emergent *Garuga floribunda* and *Albizia carrii* on the slopes. Wet Season.

Plate 3. Motupore Island, east coast with strand vegetation. ***Guettarda speciosa*** **and** ***Canthium suborbiculare*** **above the beach (to the left); mangrove with** ***Osbornia octodonta*** **and** ***Rhizophora*** **spp to seaward (to the right).**

Plate 4. Motupore Island, southeast slope with savanna of mature ***Eucalyptus confertiflora*** **above** ***Themeda triandra.***

Plate 5. Motupore Island, northeast end. Well developed monsoon scrub with emergent *Bombax ceiba* (centre) and *Albizia carrii* (right) over canopy of *Elaeocarpus arnhemicus*. Late wet season.

Plate 6. Motupore Island, northeast side. Saplings of *Bombax ceiba and Alstonia spectabilis* growing in grass at the forest-savanna boundary.

Plate 7. Motupore Island, north end. Mangrove, mostly *Rhizophora* spp to seaward (left). Behind the mown grass area of the research station is a small patch of well-developed monsoon scrub with *Eucalyptus* savanna on the hillside above.

Plate 8. Motupore Island, north summit six months after the fire in November, 1992. The central shrub is *Flueggea virosa* regenerating from the rootstock. Many of the *Eucalyptus* trees are dead while the ground flora is almost entirely composed of quick growing *Vigna radiata*.

Macroptilium atropurpureum (DC.) Urban

(Verdcourt: 505)

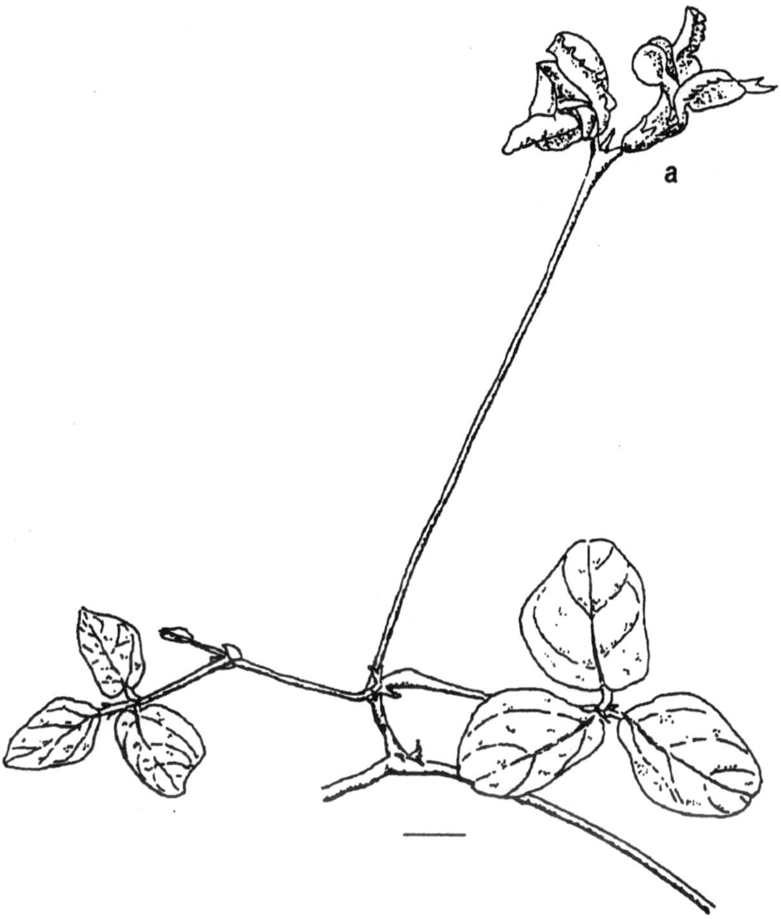

Trailing herb. Stems finely pubescent. Leaflets trifoliate; terminal leaflet broadly ovate, laterals asymmetrical with a lobe on the outer side, slightly rough, pubescent above, velutinous below. Inflorescence an axillary spike, to 25 cm long (a). Standard and keel petals pale greenish-red, keel spirally twisted, wings dark purple-black. Pods linear, c. 7 cm long, spirally dehiscent, the valves covered with silvery hairs.

Station area, near buildings.

Weed; native to tropical America, widely naturalized throughout tropics.

Frodin & Pernetta UPNG 6662; HCFH 1047/13579.

Ormocarpum orientale (Spreng.) Merr. [not illustrated]

(Verdcourt: 364)

Small shrub or tree, to 7 m high. Leaves fasciculate on short shoots, pinnate. Leaflets 9-20, alternate, oblong to oblong-ovate, usually 1.6-3 x 0.7-2 cm, rounded at apex and cuneate to rounded at base, glabrous, glaucous beneath. Flowers fasciculate on very short branchlets. Pedicels 1.5-2.5 cm long with persistent bracteoles; calyx glabrous, tube 3.5-5 mm long and teeth 4-7 mm long; corolla pale yellowish-green with veining or streaking on standard. Pods with 1-7 articles, each 1.5-2.5 cm long, 6-9 mm wide, with a corky pericarp.

Collected on the ridge at the S end in 1984 but not found since.

Thailand, China, Java, Borneo, Philippines, Sulawesi and Moluccas.

Frodin UPNG 8192.

Pycnospora lutescens (Poir.) Schindl. [not illustrated]

(Verdcourt: 413)

Herb or sub-shrub, glabrescent to velutinous, with numerous stems arising from a woody rootstock. Leaves trifoliate. Leaflets oblong, elliptic or almost round, c. 0.6-3.4 x 0.6-1.8 cm, apex rounded and mucronulate, base rounded or subcordate. Inflorescence terminal, lax, 3-7 cm long. Flowers c. 8 mm long; pedicel and calyx covered in bulbous-based hairs; corolla usually blue or dark violet. Fruits oblong, inflated, 0.7-1.2 x 0.3-0.5 cm; containing numerous seeds.

Collected on the ridge at the S end, in 1984, but not found since.

Widespread in Old World tropics from Africa to India, China and N Australia.

Frodin UPNG 8195.

Pongamia pinnata (L.) Pierre

(Verdcourt: 312)

Tree to 8 m high. Leaves imparipinnate; leaflets 5-7, mid-green, glossy above, paler below, glabrous; margin slightly wavy; venation prominent. Inflorescence an axillary raceme up to 20 cm long with up to 40 flowers (a). Buds brown; corolla bright magenta, except for white keel. Pods brown, woody, eventually dehiscing (b).

Behind beach at N end and E side, and at base of grassy slope near houses.

Tropical Asia, Malesia, Australia and Pacific.

Frodin UPNG 3162, 5205; HCFH 650/13181, 1037/13569.

May belong in genus *Millettia*.

Psoralea badocana (Blanco) Blanco

(Verdcourt: 358)

Shrubby herb to c. 1 m high. Stems, leaves, inflorescence and especially calyx covered with white woolly hairs. Leaves unifoliate, conspicuously oil-dotted, soft greyish or pale green above, much corrugated. Flowers in dense, short-stalked axillary clusters (a); flower, lateral view (b), front view (c); corolla purple, becoming blue. Fruit ellipsoid, compressed, small, enclosed within calyx (d).

Grassy slopes, on E and W sides.

E Malesia and Australia.

Frodin UPNG 3173; HCFH 628/13159, 1074/13606; s.coll. UPNG 2331.

Rhynchosia minima (L.) DC.

(Verdcourt: 553)

Climbing herb. Stems pubescent. Leaves trifoliate, occasionally 4-foliate, the leaflets ± rhomboidal, the terminal one to 1.8 x 2.0 cm, grey-green, glabrous above, paler and minutely pubescent below, densely oil-dotted. Flowers in a small, lax, axillary inflorescence (a). Standard and wings orange, keel pale green. Fruit (b) oblong, 0.6-2.5 cm long, covered with white hairs.

Savanna on ridge.

Pantropical.

Frodin UPNG 4285; Kuduk & Naoni UPNG 6110; JM 94/100.

Sesbania cannabina (Retz.) Poir.

(Verdcourt: 360)

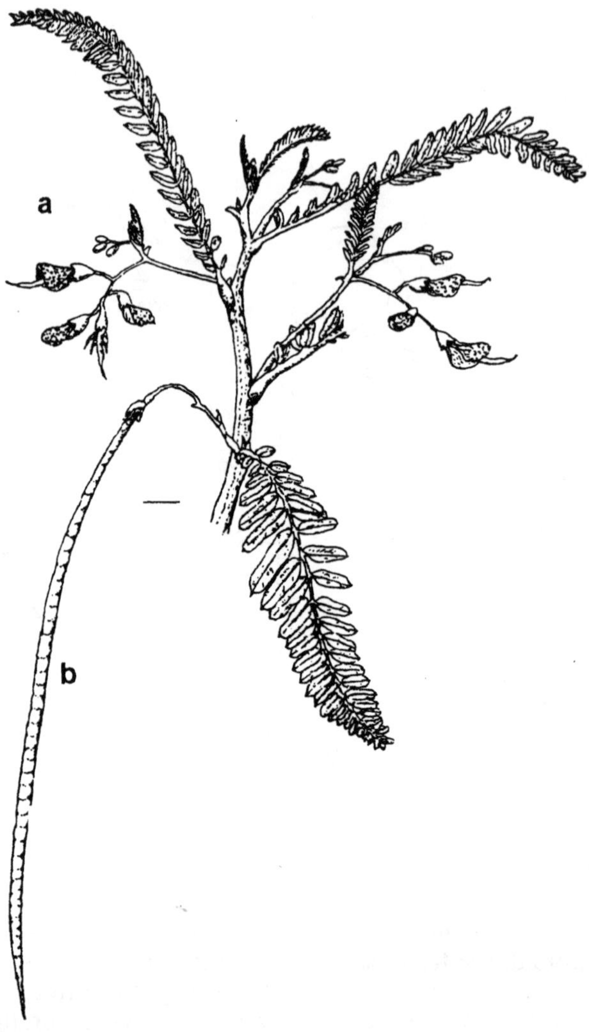

Annual little-branched herb to 1.5 m high. Stems green, glabrescent. Leaves pinnate; leaflets c. 30 pairs, c. 2 x 0.3 cm, oblong, with white scattered hairs below. Inflorescence axillary, c. 10 cm long, few-flowered (a). Corolla yellow except standard which is mottled with purple and brown on outer surface. Pods linear, glabrous, c. 15 cm long, slightly constricted between the seeds (b).

In carpet of *Ipomoea pes-caprae* at NE corner behind mangrove and beach. Not seen since 1991.

Widely distributed in tropical Africa, Asia, Malesia; probably native of Australia and Pacific.

HCFH 1048/13580.

Sophora tomentosa L.

(Verdcourt: 289)

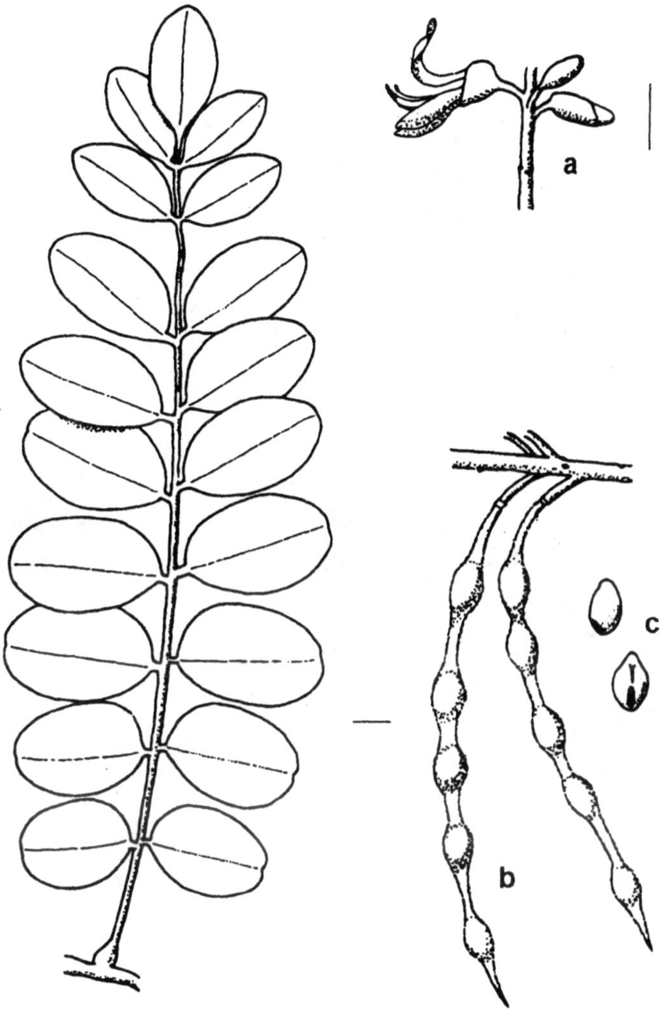

Shrub or small tree c. 3 m high with spreading branches. Whole plant has silvery green cast from pubescence on stem, leaves and fruit. Leaves imparipinnate; leaflets 9-19, upper surface glabrescent. Inflorescence a raceme, ± erect, c. 20 cm long. Flower and bud, lateral view (a); calyx silvery grey, pubescent; corolla pale yellow. Fruits moniliform (b), green turning brown. Seeds pale brown, shiny (c).

Shoreline behind mangroves at base of nature trail, E side.

Widespread in tropics and subtropics of Indo-Pacific.

Frodin UPNG 5222; HCFH 611/13142; Lambinon 87/11; Lawong s.n. 24/8/88.

Tephrosia astragaloides R.Br. ex Benth.

(Verdcourt: 340)

Shrub or woody herb, to 1 m high. Stems grooved, white from pubescence. Leaves pinnate; leaflets 7-17 pairs, oblanceolate, dull grey-green, glabrous above, densely covered with white silky hairs below. Flowers in pseudoracemes, c. 1.2 cm long, corolla cream (a). Fruit in axillary clusters of 3, spirally dehiscent, brown, the valves very shortly pubescent, 6-seeded (b).

Savanna along ridge top, N end.

N Australia and New Guinea.

Frodin UPNG 4269, s.n. April 1972; HCFH 1706/13638; Powell et al. UPNG 2328.

Tephrosia filipes Benth.

(Verdcourt: 343)

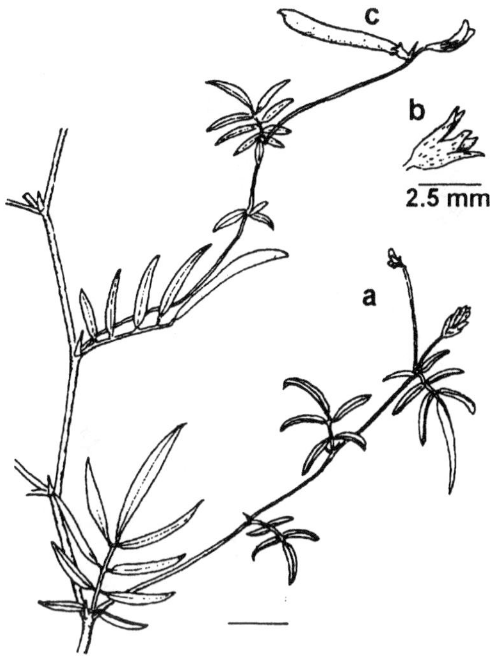

Erect or spreading much-branched herb, 25-40 cm high, with adpressed hairs on stems. Leaves pinnate; leaflets 7-15, linear-lanceolate, 0.4-3.5 cm long x 0.8-2 mm wide, usually apiculate, glabrous above and silky-pubescent below. Flowers (a) 1-2(-4) on slender peduncles 1-5 cm long; pedicels 1.5-4 mm long; calyx (b) with upper two teeth partly joined; corolla pink or purple. Pods linear, 1.5-3 cm long x 2.2-3.5 cm wide, pubescent (c).

Rare, seen only on west slope among grass.

Australia and Central Province, PNG.

Frodin UPNG 8193; JM 94/112.

Uraria lagopodoides (L.) Desv. ex DC. [not illustrated]

(Verdcourt: 415)

Ascending herb to c. 1 m high, finely pubescent and with longer hooked hairs. Leaves mostly trifoliate; leaflets elliptic, obcordate or almost round, 1.2-8 x 0.7-6.5 cm, the laterals often much smaller than the terminal one, pubescent especially below. Inflorescence sessile or shortly stalked, cylindrical, dense, spike-like, 2.5-10 cm long, bluish. Fruit folded like a concertina and enclosed within persistent calyx, consisting of 2 one-seeded inflated articles, each grey to black, shiny, 2.5-3 x 2 mm.

Collected in 1984 but not found since.

Asia, Malesia, Australia and Pacific Islands.

Frodin UPNG 8203.

Vigna radiata (L.) Wilkczek var. **sublobata** (Roxb.) Verdc.

(Verdcourt: 523)

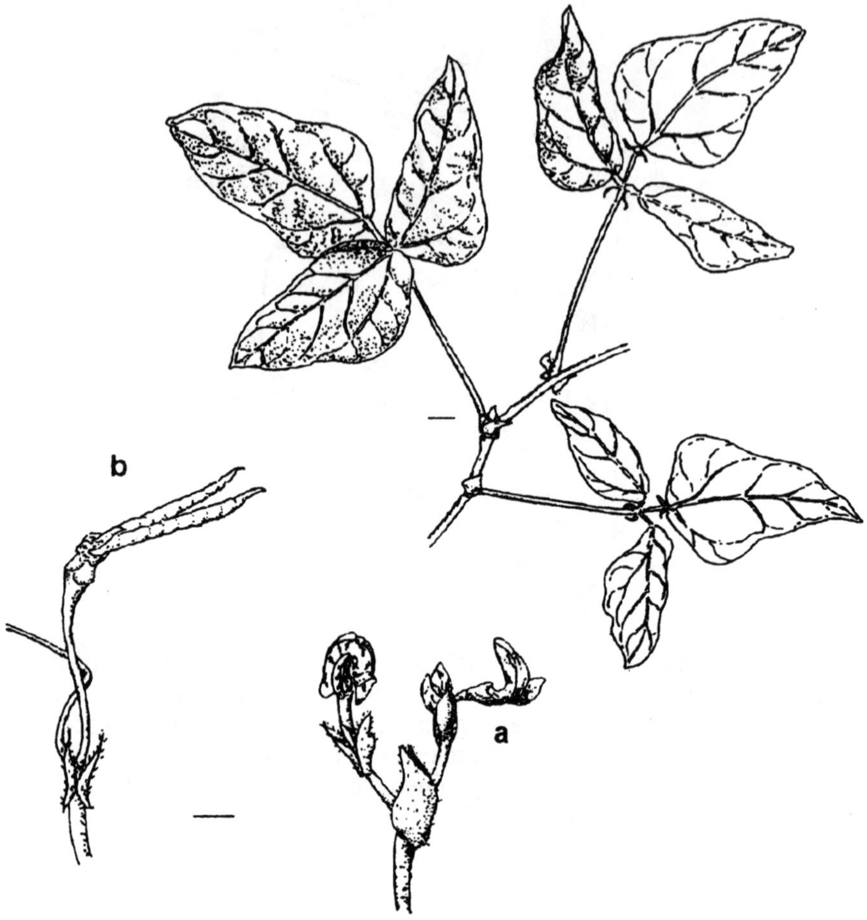

Trailing herb. Stems with white or brown hairs. Leaves trifoliate; leaflets ovate or rhomboidal, to c. 8 x 5.5 cm, the laterals asymmetrical; margin sinuate or lobed; scattered white hairs on upper and lower surfaces, somewhat rough. Inflorescence an axillary raceme (a), the flowers clustered towards the apex. Corolla greenish yellow outside; keel curved through 180°. Fruits linear-cylindrical, 4-9 cm long, somewhat constricted between seeds (b).

Savanna, common.

Widespread in Africa and Asia, often cultivated and naturalized.

HCFH 1798/13750.

MALPIGIACEAE

Rhyssopteris timoriensis (DC.) Juss. in Deless.

(Fl. Mal. 5: 140)

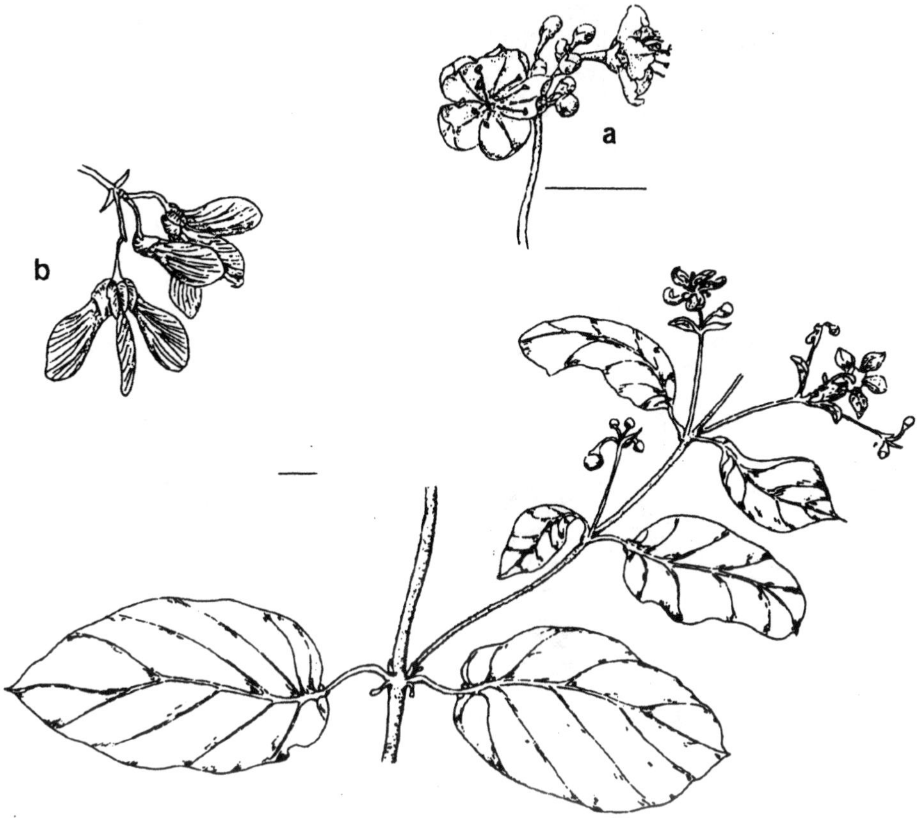

Robust climber. Stems brown with adpressed minute pubescence. Stipules 2 per leaf, sometimes with 2 additional smaller ones, obovate. Leaves simple, opposite, ovate, dull olive-green and glabrous above, minutely pubescent below; margins entire, somewhat undulate; gland on either side of petiole at junction with lamina. Flowers in axillary, pedunculate, umbel-like clusters (a). Sepals brownish yellow, small; corolla pale yellow, larger; stamens 10; styles and stigmas 4. Fruit (1-2-) 3-winged, dry, brown (b).

Frequent in savanna and at scrub edge.

Taiwan, Micronesia, Melanesia and N Australia.

HCFH 1057/13589, 2011/13763; Kuduk & Naoni UPNG 6117A; Leach UPNG 5273.

Abutilon indicum (L.) Sweet

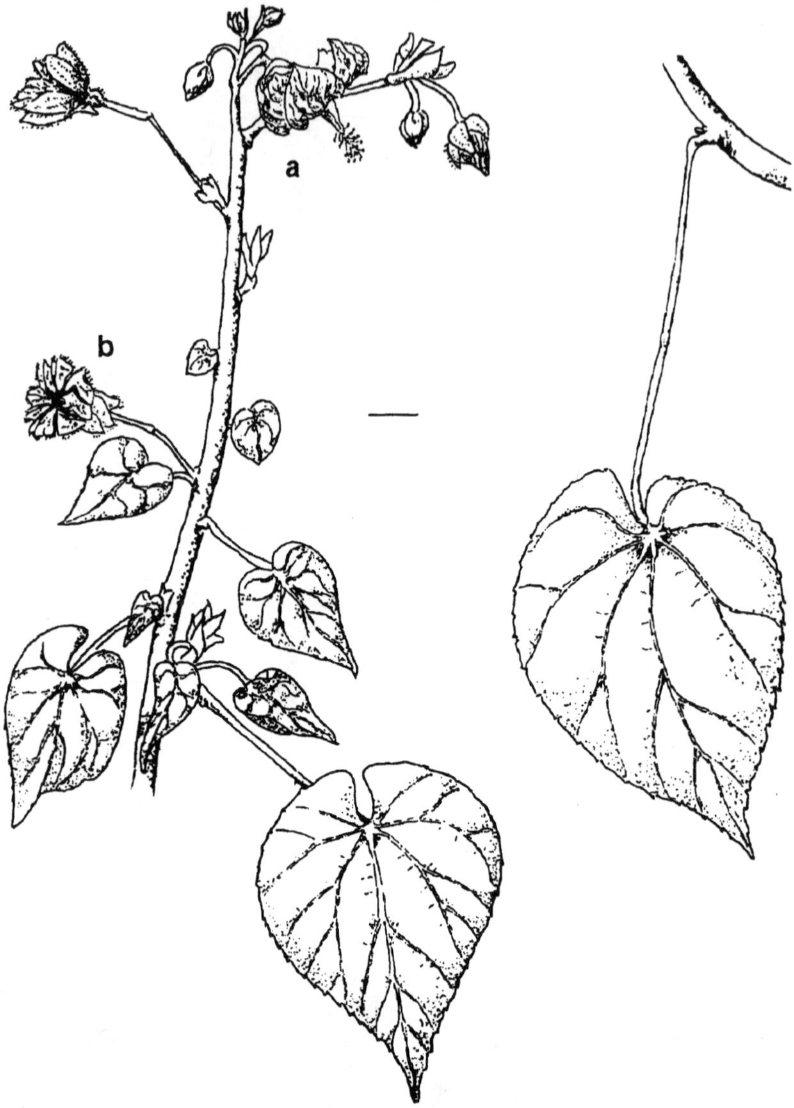

Woody herb, to 2 m high. Whole plant covered with short, softly velvety pubescence. Leaves alternate, cordate, yellow-green to grey-green, sometimes mottled above, grey-green to almost white below; margin crenulate. Flowers towards apex of leafy stems (a); corolla yellow, more orange in bud. Fruit a schizocarp (b) with numerous lines of dehiscence; valves black, densely covered with long white hairs.

Strand and coastal scrub on E and W sides, 1 m ASL, in sun.

Widely distributed in tropics and subtropics of Old and New Worlds.

Frodin UPNG 4254, 5220; HCFH 924/13455, 1750/13682; Powell et al. UPNG 2344.

Hibiscus tiliaceous L.

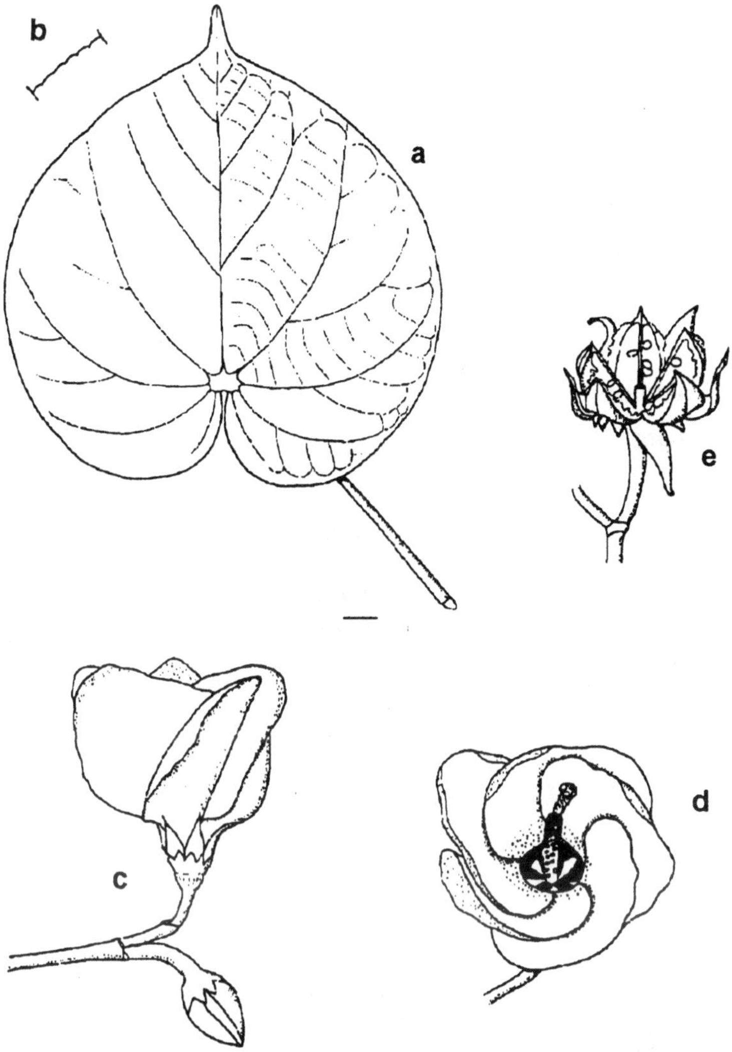

Bushy tree, to 7 m high. Young stems, petioles and pedicels grey-green and covered with very short, velvety pubescence; older stems brown with circular scar at each node. Stipules large. Leaves alternate, ± circular, deeply cordate, darkish or mid-green, glabrescent above, pale grey-green pubescent below (a); margin crenulate (b); blade perpendicular to petiole. Flowers towards apex of leaf-bearing twigs (c & d). Calyx and epicalyx grey-green pubescent; petals yellow tinged with pink, turning pink and falling on second day; base of petals and stigma deep dark red; anthers yellow. Fruit (e) a dry capsule with 5 lines of dehiscence, the valves golden brown, pubescent; seeds numerous.

Near Old Hostel on flat open ground, and strand near houses.

Circumtropical, especially on coasts; seeds distributed by ocean currents.

Frodin UPNG 3176; HCFH 1082/13614.

Hibiscus schizopetalus (Mast.) Hook.f. [not illustrated]

Shrub to 2 m tall. Leaves simple, alternate, glabrous, pale green-yellow; margin serrate. Flowers red; corolla deeply dissected. Fruit never found in Malesia.

Cultivated near Old Hostel.

Native to E Africa; elsewhere as an ornamental.

Frodin UPNG 3196.

Malvastrum coromandelianum (L.) Garcke [not illustrated]

(Henty & Pritchard: 124)

Tough herb with woody rootstock, to 20 cm high. Stems sometimes reddish, covered with long white hairs when young, glabrescent towards base. Leaves alternate, ovate, shortly petiolate, corrugated, with long pale hairs especially along veins on lower side; hairs fewer above; margin serrate. Flowers axillary; epicalyx of 3 small segments; calyx 5-lobed; corolla yellow; fruit with 10 one-seeded segments (mericarps) each c. 3 mm long.

Flat open ground, Station area, in full sun.

Pantropical weed with preference for seasonal climates.

Frodin & Pernetta UPNG 6645; HCFH 1776/13708.

Sida cordifolia L. [not illustrated]

(Henty & Pritchard: 12)

Erect herb to 0.5 m high. Stems green, covered in soft, short, woolly, yellowish hairs. Leaves simple, alternate, ovate, softly hairy on both sides; the petiole about equal in length to the blade; margin dentate. Flowers in axillary clusters, yellow; calyx densely hairy. Fruit dividing into 10 mericarps.

In shade between flat grassy area and scrub, N end.

Pantropical.

HCFH 1071/13633.

Thespesia populnea (L.) Sol. ex Correa

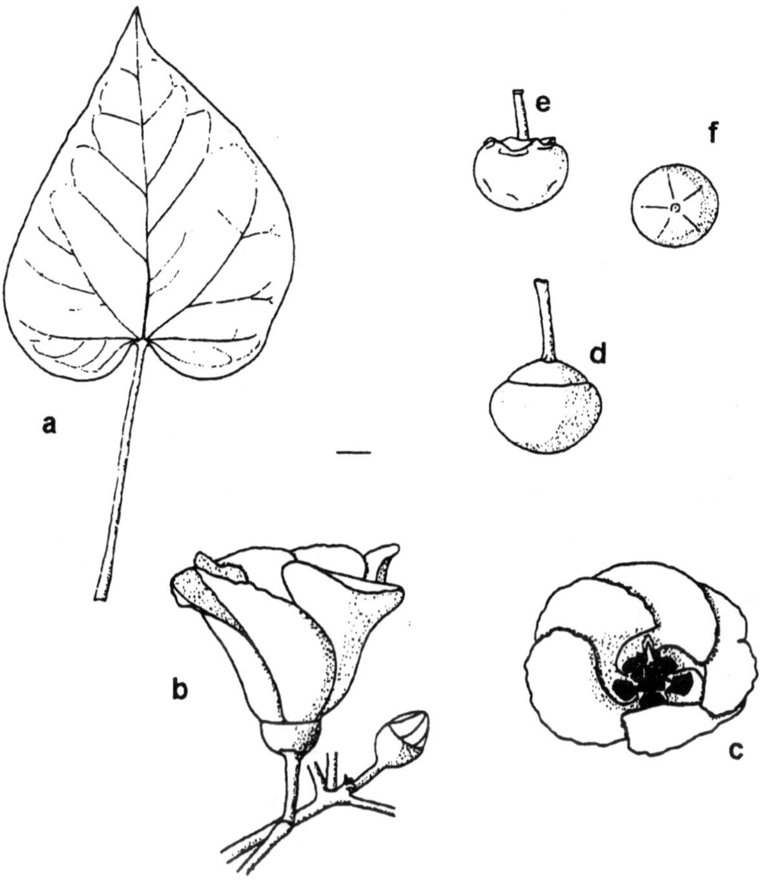

Tree to 8 m tall. Stems, petioles and pedicels brownish or green, with minute waxy scales. Leaves simple, alternate, triangular or somewhat cordate (a), darkish green, glabrous above with yellow venation, glabrous below; margin entire though often misshapen and insect-damaged; petiole not perpendicular to blade. Flowers (b and c) towards apex of leaf-bearing twigs. Calyx entire, minutely hairy; corolla pale yellow with 5 red spots inside at base, turning purple at tip and salmon pink below on second day; androecium yellow; stigmas fused, yellow, palest at tip (c); sweetly scented. Immature fruit hard, green (d). Mature fruit (lateral view, e; end view, f) very dark brown, ± shrunken, indehiscent but with 5 lines of weakness, often falling whole; calyx persistent, shrivelled and undulating.

Very common along strand line and behind mangroves.

Widely distributed in tropics; in Malesia, confined to sea shore.

Frodin UPNG 5205; HCFH 1085/13617.

Turraea virens L.

T. pubescens J. Hell.

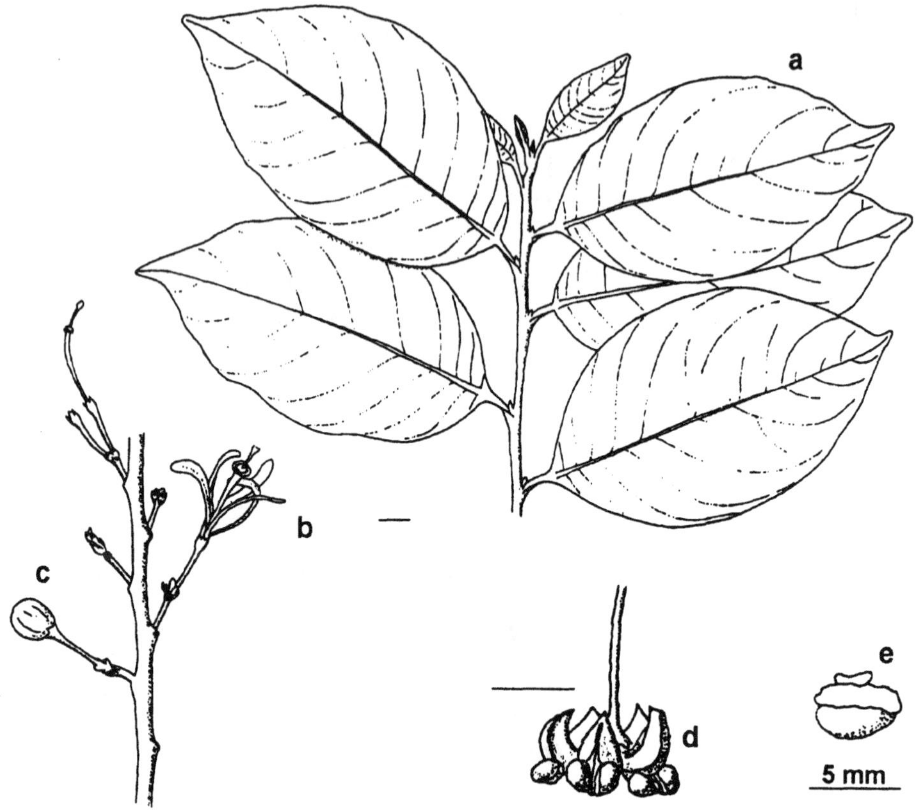

Straggly shrub or small tree to 4 m high; branches often horizontal. Stems brown, woody, glabrous; young twigs minutely hairy. Leaves alternate, distichous (a), yellow-green and softly velvety on both surfaces, though may be less velvety above; margin entire, ± undulating; venation yellowish. Flowers on short side shoots (b) arising from axillary buds, often on lower, leafless, part of twigs. Flowers strongly scented; calyx green; corolla, staminal tube and style white changing to pale yellow; anthers yellow, stigma box-like, greenish. Fruit a green, ribbed capsule (c), with 5 sutures; valves reflexed after dehiscence (d), cream or orange within; seeds 10, pendulous, black with orange aril (e).

Common in savanna and scrub margins.

India, Thailand, Cambodia, Hainan, Java E to Philippines, New Guinea and N Australia.

Frodin UPNG 842, 4256, 5214; HCFH 607/13138, 1737/13669; Huxley s.n. 15/6/75; Kuduk & Naoni UPNG 6123; Leach s.n. Oct 1979; Naoni & Kuduk s.n. 29/1/83; Powell et al. UPNG 2346.

Xylocarpus rumphii (Kostel.) Mabb.
X. moluccensis (Lam.) M. Roem.

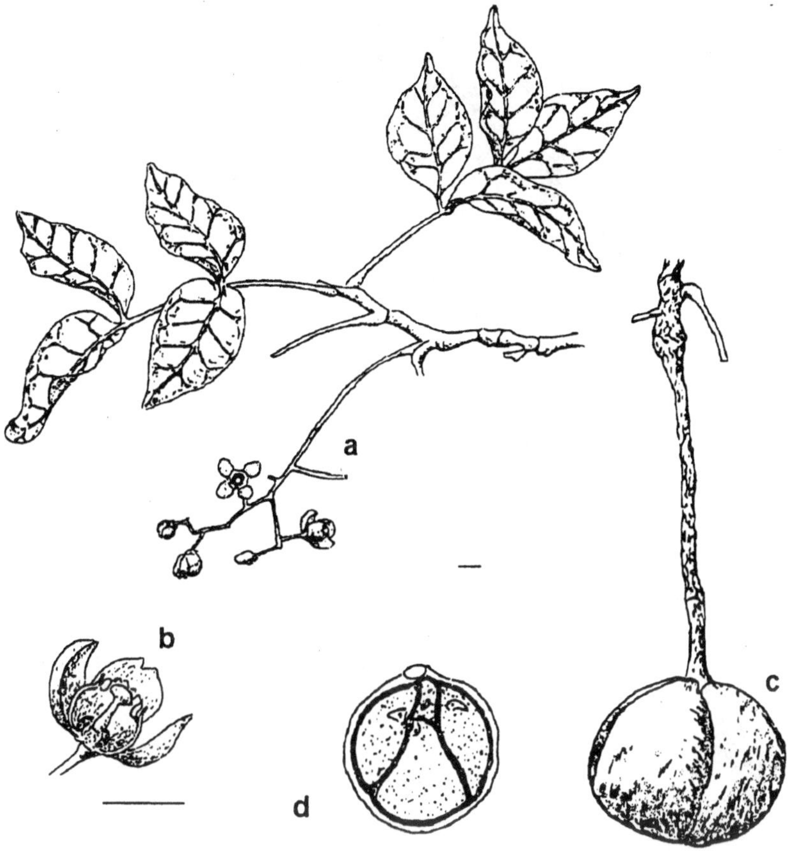

Spreading tree to 10 m high with thick bark. Stems brown, smooth when young. Leaves paripinnate; leaflets in 2 pairs, opposite, ± symetrical, mid-green, glossy above, paler below, glabrous; margins entire. Inflorescence pendent, lax (a). Flowers creamy white; stamens 6, inside a flask-like structure (b). Fruit green, mottled, spherical (c), containing 8 pale brown seeds which fit together like a puzzle (logitudinal section, d).

Occasional, strand, E and W sides.

Widespread in Malesia and beyond.

Frodin UPNG 847; HCFH 1096/13628; Leach UPNG 5272; Powell et al. UPNG 2351; Totome RGT 25.

Xylocarpus granatum Koenig, one of two mangrove species in this genus, was recorded by Frodin (no date) as a new arrival on Motupore in 1985 but we did not see it in 1990. It has a short, upright (not pendulous) inflorescence and spherical fruit up to 12 cm in diameter. The other mangrove species, *X. australasicus* Ridley, is common on Bootless Bay but never recorded from Motupore. It also has a short, upright inflorescence. (See Frodin et al. 1982: 70.)

Fatoua pilosa Gaudich. in Freyc.

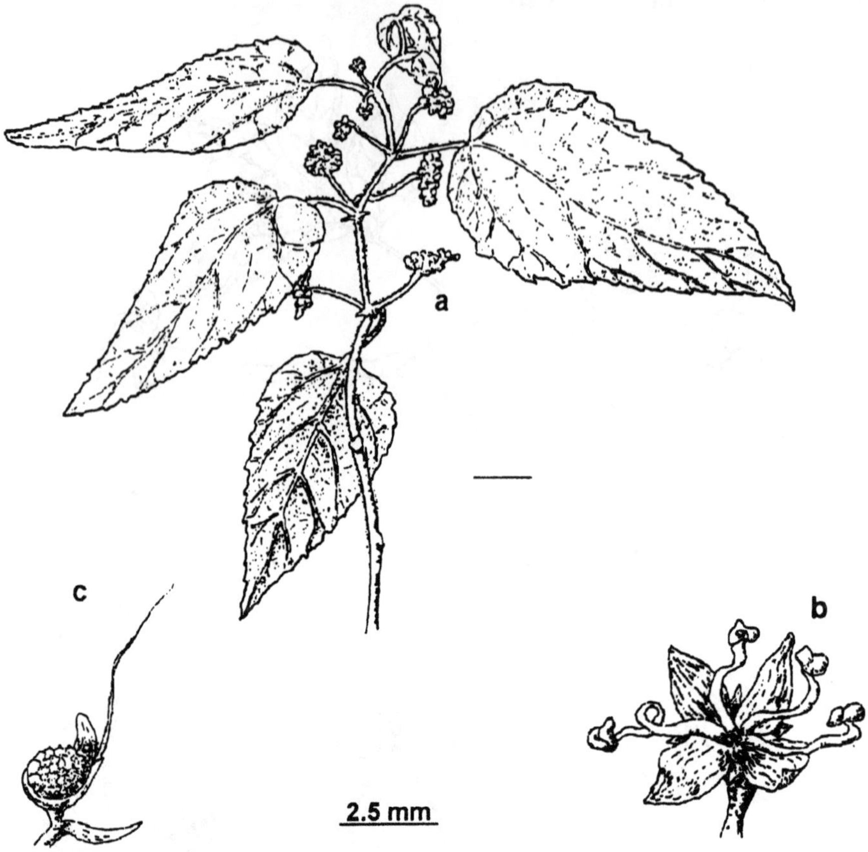

Lignified herb, c. 0.5 (-1) m high. Leaves simple, alternate, only towards top of upright woody stems, dull mid-green, roughly hairy on both sides; margin crenate to dentate. Flowers monoecious, tiny, in pedunculate axillary clusters (a), yellowish; male flowers with white, ribbon-like filaments (b); female flowers with white ovary (c).

Common in patches in dense shade in scrub.

Java to Japan and N Australia.

HCFH 1098/13630; Kuduk & Naoni UPNG 6112; Lambinon 87/92.

Ficus microcarpa L.f. var. **rigo** (Bailey) Corner

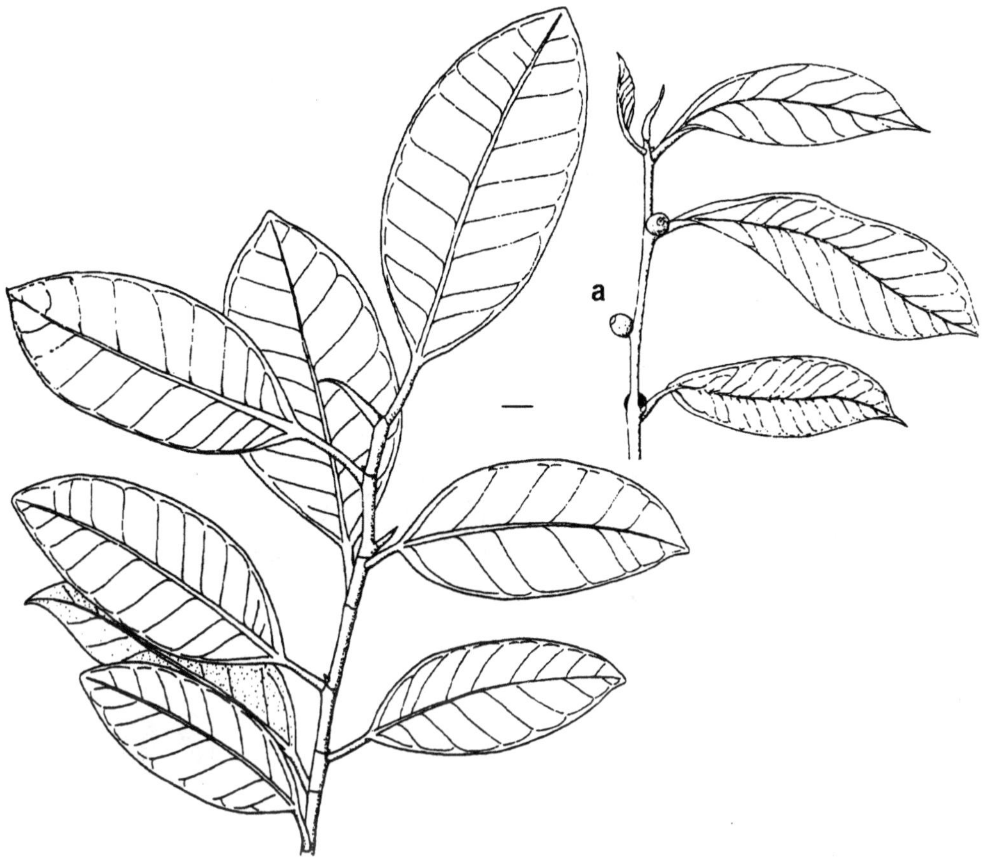

Strangling tree, to 5+ m. Sap milky. Twigs smooth, grey-brown. Leaves simple, alternate, glabrous, dark glossy green above, mid-green below, drying glaucous; petiole yellow-green; venation pale green-yellow, distinct above and obscure below; margin entire. Figs (a) sessile, axillary, with several basal bracts, c. 1 cm diameter, green turning orange.

Several in scrub on W side and growing out of rocks above strand line.

Ficus microcarpa occurs from India and Sri Lanka, S China, through SE Asia, Malesia and Australia, to Caroline Islands. Var. *rigo* is endemic to Central Province, Papua New Guinea; according to D. Frodin (pers. comm.) it is worthy of specific rank.

HCFH 1066/13598; JM 92/19.

Ficus microcarpa var. **latifolia** (Miq.) Corner is also recorded from Motupore by Frodin (no date).

Ficus opposita Miq.

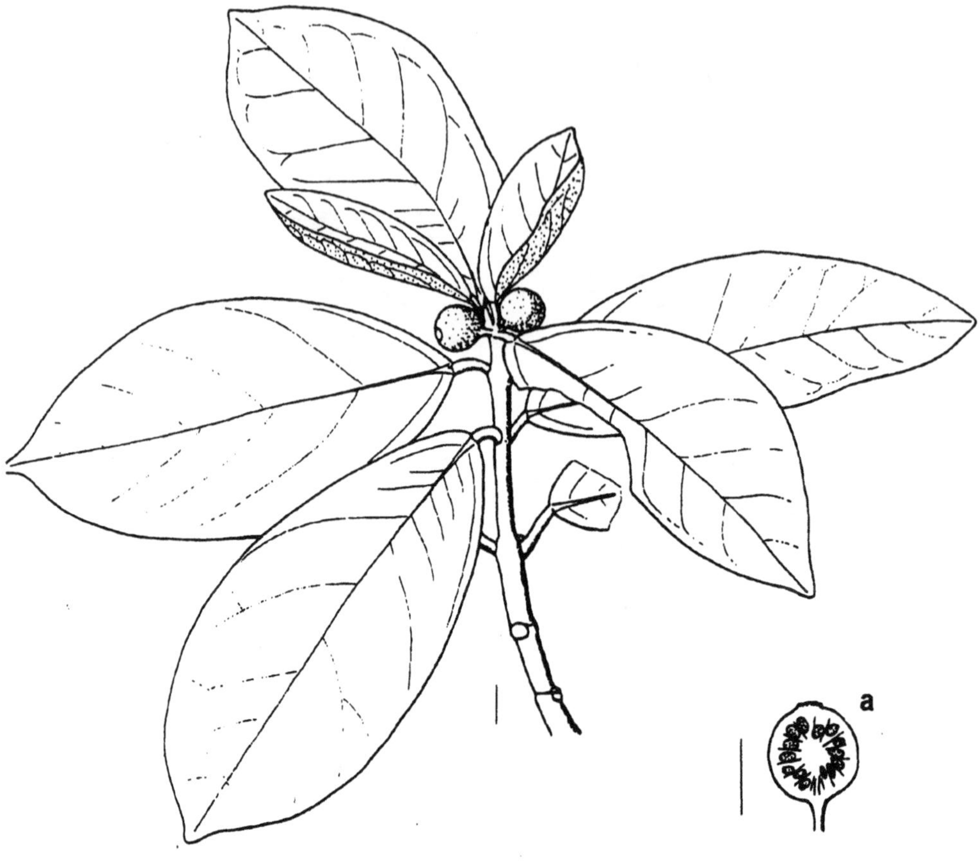

Shrub or less commonly a small tree, 2-8 m high with milky sap. Young twigs rough-hairy, old ones smooth. Leaves simple, mostly opposite, very variable in size from c. 6.5 x 3.5 to 17 x 9 cm, rough-hairy, like sandpaper, on both sides, light or mid-green above; venation distinct; margin entire. Leaves mostly elliptical but in *Frodin s.n., 30-31 March 1974*, which has opposite leaves, they are deeply 3-lobed, the central lobe being long and thin. *HCFH 1758/13690*, from a sapling in scrub near the Caretaker's house, with narrowly elliptical leaves, may be a juvenile form. Figs axillary, on very short peduncles (longitudinal section, a), skin green ripening deep wine red, the flesh white, 'seeds' orange.

Very common in savanna.

Java, New Guinea and Australia.

Frodin et al. UPNG 2928, s.n. 1972, s.n. 30-31 March 1974; HCFH 1051/13583, 1707/13639; Kuduk & Naoni UPNG 6119.

Maclura cochinchinensis (Lour.) Corner

Cudrania javanensis Tréc.

Trailing liane with milky latex. Stems cane-like, scurfy, glaucous-green, with white lenticels. Leaves simple, alternate, variable in size from 2 x 1 to 7 x 3 cm, glabrous, mid-green above, much paler below; margin entire; single sharp thorn in each leaf axil, at least of larger leaves. Dioecious. Flowers in globose axillary heads, 4-partite. Composite fruit globose, 2-3 cm diameter, tesselated, ripening bright orange (a).

Rare on scrub margin on E hillside, behind Caretaker's house.

East Africa, India and Sri Lanka, Indo-China, Malesia and Australia.

-Frodin UPNG 3186 (small leaves, side shoots); HCFH 1709/13641 (larger leaves; illustrated, primary canes).

"The correct name for our plant is uncertain, but if two species are recognized [it is] *Maclura amboinensis* Blume. In the Trobriands is another species which may be the true *cochinchinensis*. Clarification is required." (Checklist of Port Moresby Region (card index, UPNG herbarium), D. Frodin, unpublished).

Trophis scandens (Lour.) Hook. & Arn.

Malaisia scandens (Lour.) Planch.

Woody or wiry scandent liane with white sap, at least in young shoots. Stems rough, like sandpaper, brown with white lenticels. Leaves simple, alternate, either elliptical with slightly sinuate margin (a), or asymmetrically lobed (juvenile form?) and borne on side shoots of main stem (b), rough, glabrous, slightly shiny green above; venation yellowish, prominent below. Flowers minute; female flower (c). Infructescence branched; immature fruit green (d); mature fruit red, fleshy, several embedded on a warty base that is green initially, turning pink (e).

Savanna and scrub on E and W sides.

SE Asia to Australia, New Guinea and New Caledonia.

Frodin UPNG 3180, 5248; HCFH 1714/13646, 1760/13692; JM 91/9.

MYRSINACEAE

Aegiceras corniculatum (L.) Blanco [not illustrated]

(Frodin et al: 14)

Small bushy tree, c. 7 m tall. Twigs brown, smooth. Leaves simple, alternate, elliptical, to c. 8 x 3.5 cm, ± coriaceous, glabrous, yellow-green above, paler below; margin entire; petiole c. 1 cm. Flowers in a pedicelate cluster, white. Large buds pointed at apex. Fruits cylindrical, c. 3-4 x 0.4 cm, strongly curved or horn-shaped, pointed at apex.

Occasional in mangrove on E side and on landward side on gravelly beach at NW corner .

S and SE Asia, Malesia to Australia.

Frodin UPNG 814; HCFH 1089/13621; Powell UPNG 2322.

Trophis scandens (Lour.) Hook. & Arn.

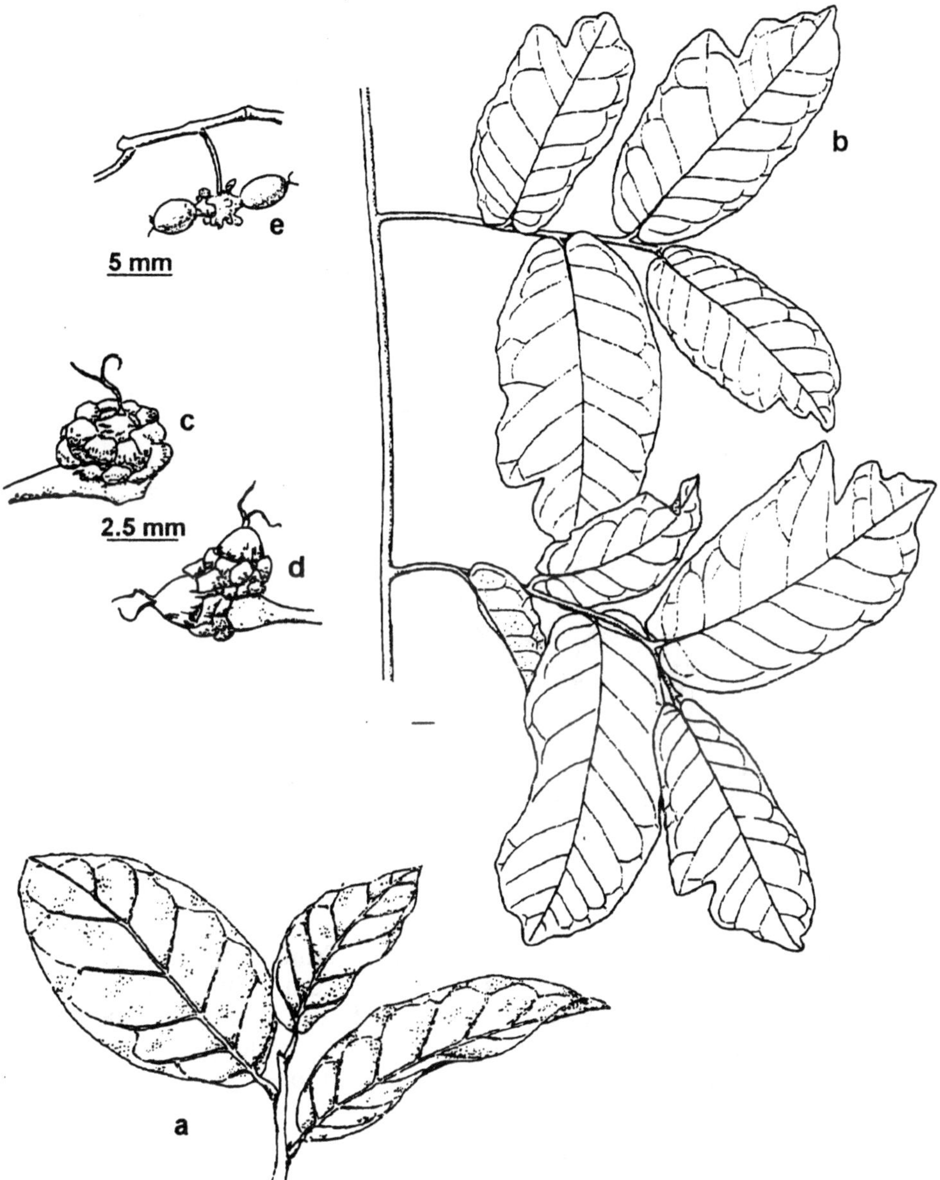

Eucalyptus species

Only two species of *Eucalyptus* (*E. confertiflora* and *E. papuana*) are recorded from Motupore. However, many trees show intermediate characters, probably due to hybridisation and introgression, and a single tree may show a variety of leaf morphologies, as shown for *E. papuana*. *E. confertiflora* on Motupore is unusual in that the leaves are nearly glabrous, like those of *E. papuana*, whereas the usual form has pubescent leaves.

Eucalyptus confertiflora F. Muell.

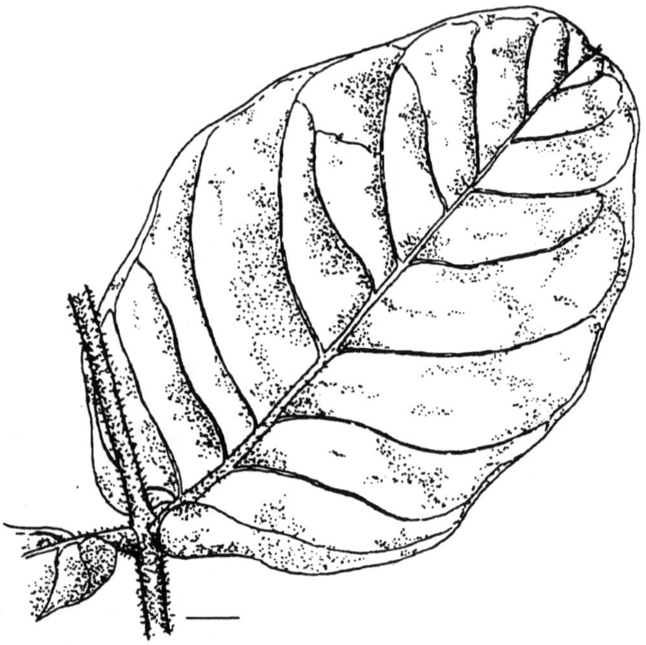

Crooked or straggly tree, to 10 m high. Trunk smooth with exfoliating bark and producing large quantities of sticky gum. Branches and twigs ± ascending. Leaves simple, alternate, hard and stiff, mostly isobilateral, grey-green on both sides, nearly glabrous, narrowly to broadly ovate or elliptical, cordate at base, with oil dots. Flowers and fruit as for *E. papuana*.

Dominant tree in savanna.

N Australia and New Guinea.

Frodin et al. UPNG 2930, 4270; Lambinon 87/107.

Eucalyptus papuana F. Muell.

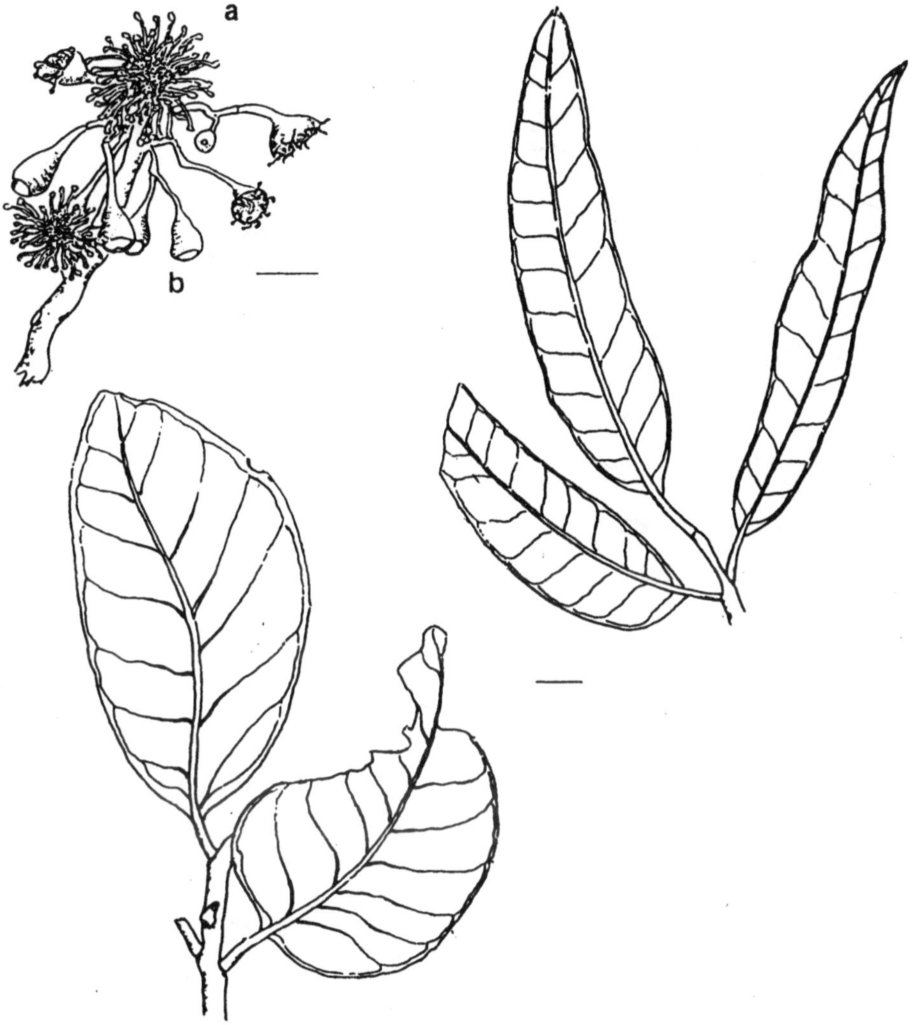

Tree to c. 10 m high. Trunk smooth with exfoliating bark and producing sticky gum. Twigs and foliage drooping or trailing, giving characteristic "weeping" appearance. Leaves simple, alternate, isobilateral, grey-green on both sides, glabrous, lanceolate, obtuse or rounded at base, hard and stiff, with oil dots. Flowers in clusters (a), the petals forming a rounded or flattish cap which falls as the flowers open; stamens numerous, white. Fruits hard, dry, woody, thin-walled, papery, cup-shaped (b).

Occasional in savanna.

Australia and New Guinea.

HCFH 1708A/13640A, 2015/13767.

Jossinia desmantha (Blume) Blume

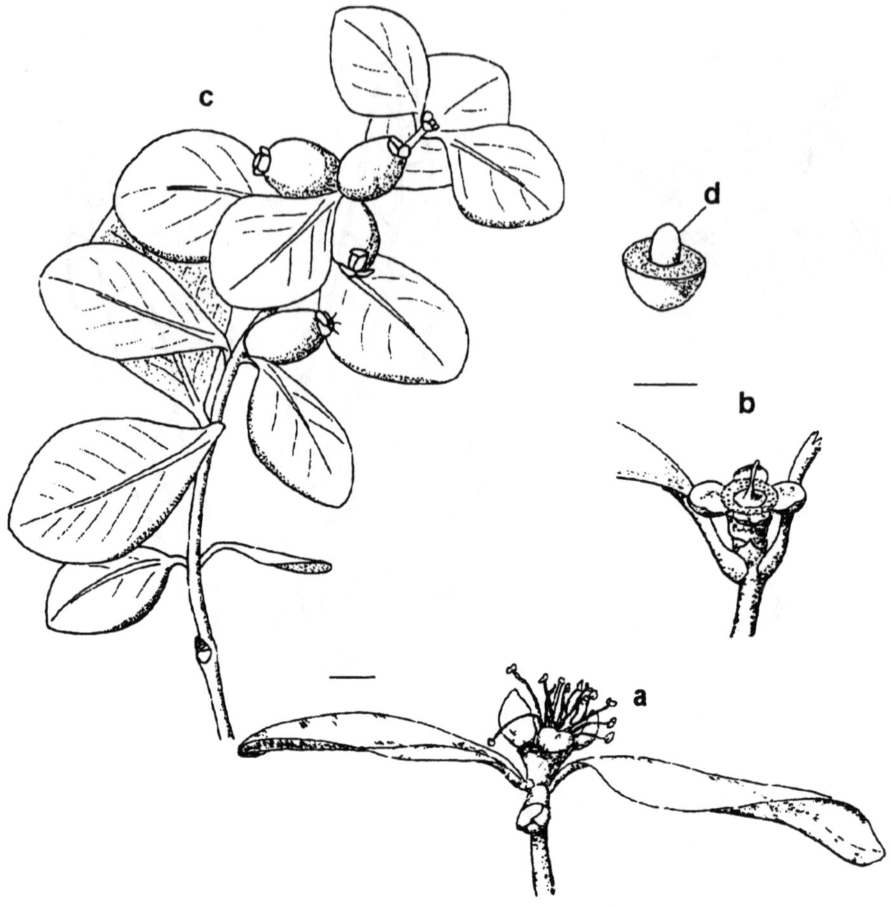

Shrub 2-5 m high. Twigs brown with fine longitudinal fissures in bark. Leaves simple, opposite, thick and stiff, mid-green or yellow-green above, glabrous, finely pitted with oil dots above; venation faint above and obscure below; margin entire; faintly fragrant when crushed. Flowers single or in pairs in leaf axils (a); calyx of 4 lobes; petals 4, white; stamens numerous, white. Immature fruits globose, calyx lobes and a few stamens persisting (b). Mature fruit (c); exocarp yellow, turning dark purple-red with purple-staining flesh and containing a single pale seed (d).

Strand line, behind mangroves on E side and at scrub edge on W side.

Endemic to the Port Moresby area.

Frodin UPNG 812, 4487; HCFH 1071/13603.

Probably closely related to *Jossinia reinwardtiana* (Blume) Blume which occurs from Borneo east to Polynesia and Hawaii; sometimes placed in the genus *Eugenia.*

Osbornia octodonta F. Muell. [not illustrated]

(Frodin et al: 48)

Bushy much branched, rounded tree to 6 m high. Twigs 4-angled, sometimes reddish. Leaves simple, opposite, obovate, notched at apex, to 5 x 1.8 cm, glabrous, oil-dotted, bright mid-green to dull grey-green, drying brown, with distinctive red petioles; margin crenulate; pleasant 'eucalyptus oil' smell when crushed. Flowers 1 (-3) in leaf axils, sessile; calyx and inferior ovary covered with grey-white hairs, 8-lobed. Fruit small, dry, cone-shaped.

Mangrove, NE corner near sand spit and along E side.

Borneo, Java, Philippines to New Guinea and N Australia.

Frodin UPNG 817, 5242; HCFH 612/13143; Lambinon 87/102; Powell et al. UPNG 2332.

NYCTAGINACEAE

Boerhavia diffusa L.

? *B. repens* L.

(Fl. Mal. **6**: 454; Henty & Pritchard: 130)

Herb with slender ascending, often reddish stems arising from a woody rootstock. Leaves simple, opposite, of unequal size within each pair, the larger 2-4 x 0.8-2.4 cm, glabrous or sparsely hairy, green suffused with purple; margin wavy; petiole short; upper leaves sessile. Flowers small, in axillary panicles (a); perianth 5-lobed, pink. Fruit 5-ribbed, cylindrical, 4 mm long.

Open ground in Station area and near ridge top by trail.

Pantropical.

Frodin et al. UPNG 5235 ***(repens)****, 5236* ***(diffusa)****, 6648; Lambinon 87/90; Leach UPNG 5275.*

Flora Malesiana gives *B. repens* as a synonym of *B. diffusa*, although Fosberg (Smithsonian Contrib. Bot. **39**. 1978) regards them as separate. *B. repens* is often distinguished from *B. diffusa* by green, not purple-tinged, leaves, and white, not pink, flowers.

Boerhavia erecta L. (Henty & Pritchard: 131), another pantropical weed, is a similar but more robust plant with almost fleshy leaves and stems. The flowers (b) are bright magentaFound on bare ground at E foot of nature trail, in shade at scrub edge. *JM 92/30.*

Boerhavia diffusa L.

Boerhavia erecta L.

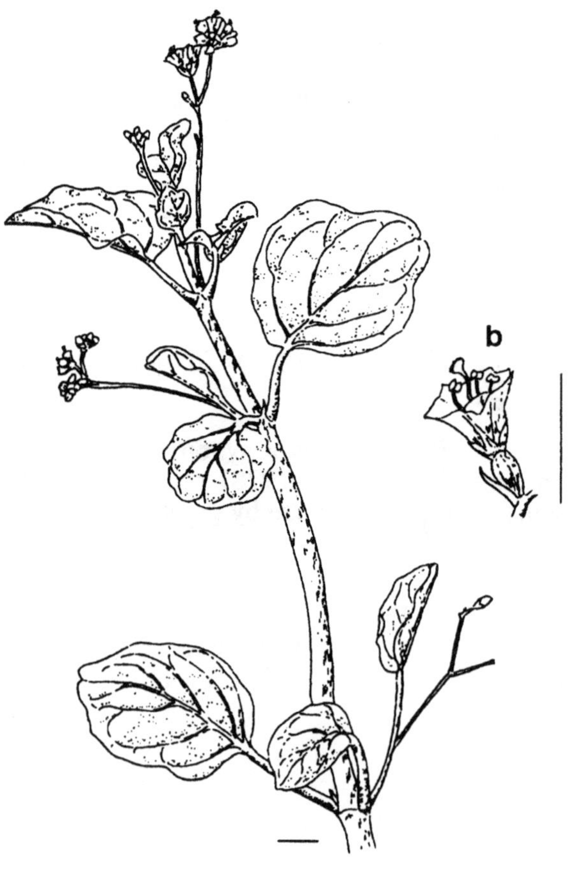

Chionanthus ramiflorus Roxb.

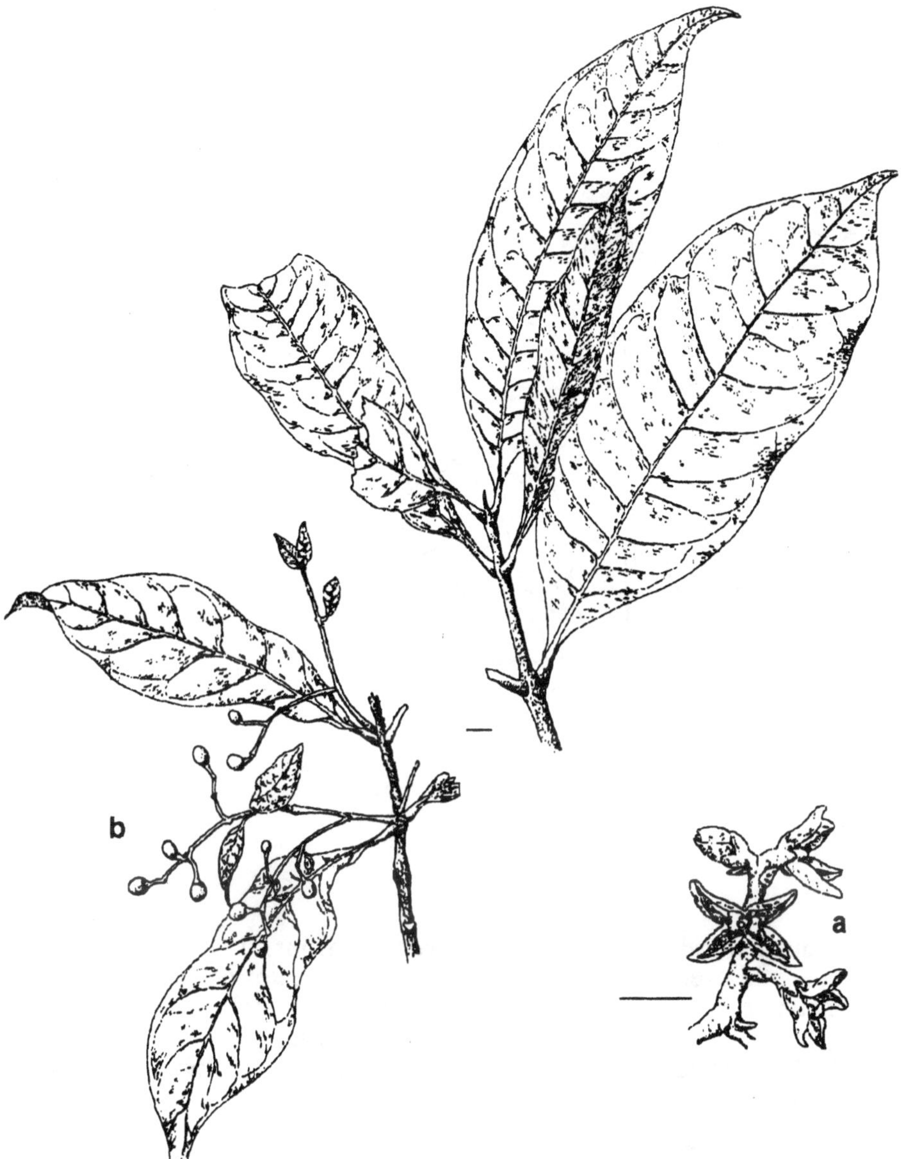

Small, single-stemmed tree, to 6 m. Leaves opposite and decussate, entirely glabrous, ovate. Inflorescence axillary, branched, bearing small leaves. Flowers yellow; petals 4, thick, rolled inwards at the edges (a). Fruits ovoid (b), green, turning black, 1 or 2-seeded.

Single tree on the west slope, at grass-scrub boundary.

Sumatra E to New Guinea.

JM 91/3, 91/16.

Jasminum aemulum R.Br.

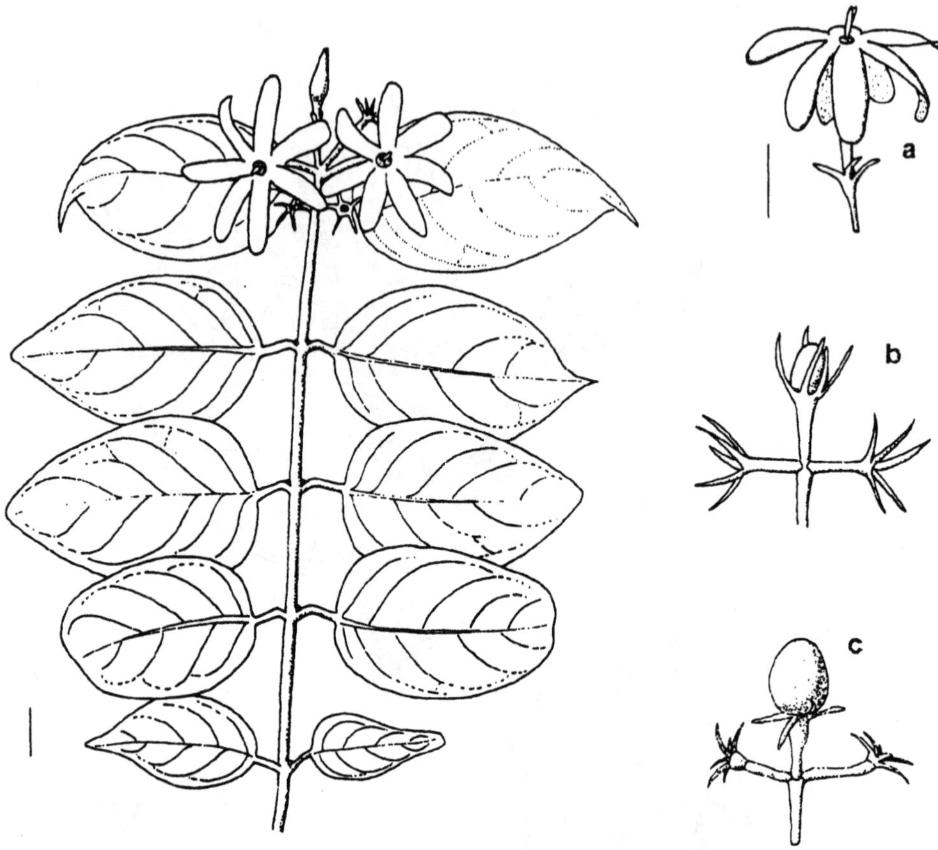

Trailing woody climber. Stems green, glabrous. Leaves simple, opposite, distichous, mid-green and sometimes shiny above, paler below, usually glabrous but sometimes sparsely hairy above and pubescent below; venation not very distinct; margin entire. Flowers in cymes, fragrant (a). Calyx with narrow, pointed lobes, persistent in fruit (b); corolla white, 5- to 7-lobed. Fruit ellipsoid, superior, fleshy, green when immature turning shiny black (c).

Abundant in savanna and at edge of scrub.

Australia, New Guinea (and ? Philippines).

Frodin UPNG 4250; HCFH 618/13149, 657/13188, 1726/13658; Huxley s.n. 15/6/75; Kuduk & Naoni UPNG 6117B.

Jasminum didymum Forst. f.

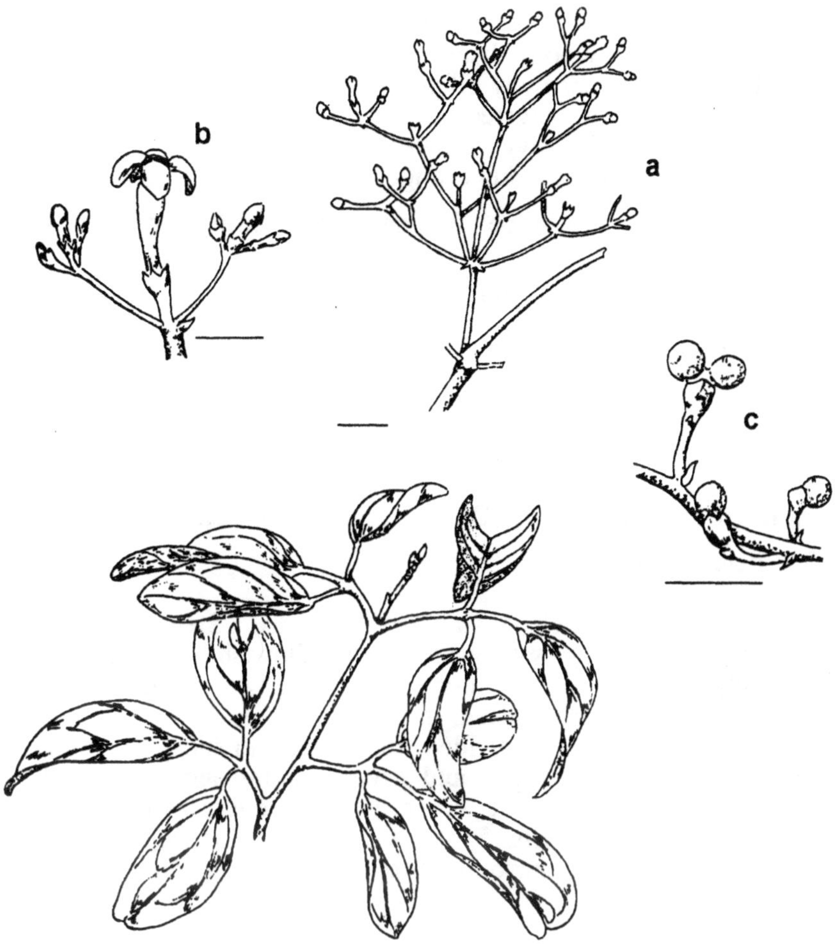

Woody trailing climber. Stems dull green to brown. Leaves usually opposite, trifoliate, the terminal leaflet usually largest; leaflets ovate, acute or notched at apex, glossy yellow-green above, paler below, glabrous; margin entire; sometimes conduplicate. Flowers in compound cymes (a). Calyx lobes very short; corolla tubular, white, though greenish towards base, 4 to 5-lobed, fragrant (b). Immature fruit (c) translucent green, single or double, maturing dull black, spherical, fleshy, c. 1 cm diameter, one seed per lobe.

Common in savanna and at edge of scrub.

(Java?), Philippines to Australia, New Caledonia and Polynesia.

Frodin UPNG 3170; HCFH 658/13189, 1059/13591, 1725/13657; Huxley s.n. 15 & 16/6/75; Leach UPNG 5271.

OPILIACEAE

Opilia amentacea Roxb. [not illustrated]

(Fl. Mal **10**: 47)

Large, scrambling woody liane. Twigs slightly thickened at nodes. Leaves simple, alternate, narrowly elliptical, c. 8 x 2.6 cm, acute at apex and attenuate at base (though reported as very variable), coriaceous, glabrous, bright green; margin slightly wavy. Flowers small, in axillary racemes; tepals 4-5, free, greenish yellow. Fruit a dull orange, fleshy drupe. A root parasite.

Rare in scrub on steep slope, W side. Reported by Frodin as seen only once. Not found in 1990.

Tropical Africa, S Asia, Malesia to Solomon Islands and Australia.

Frodin UPNG 2941.

PASSIFLORACEAE

Passiflora foetida L. var **hispida** (DC. ex Triana & Planch.) Killip

(Henty & Pritchard: 136)

Herbaceous, foetid-smelling vine with tendrils. Stem, leaves, petioles, pedicels etc. covered in long, soft, greenish or yellowish hairs. Leaves simple, 3-lobed, spirally arranged, yellowish-green, palmately nerved; margin bearing glandular hairs. Tendrils unbranched, in leaf axils. Flowers (a) solitary in leaf axils; bracts 3, finely and deeply divided, terminating in sticky glandular points; calyx 5-lobed, with white margins and green central nerve; petals 5, white; corona filamentous, purple at base, cream at apex; ovary and stamens form a central androgynophore; stamens 5; styles 3. Fruit globose, green, maturing orange; exocarp coriaceous, surrounded by persistent, green, lacy bracts (b); pulp inside like cotton wool.

Widespread, except in scrub.

Weed; native of S America, established elsewhere in tropics, widespread in SW Pacific.

HCFH 2016/13768; Huxley s.n. 15/6/75.

Passiflora foetida L. var. **hispida** (DC. ex Triana & Planch.) Killip

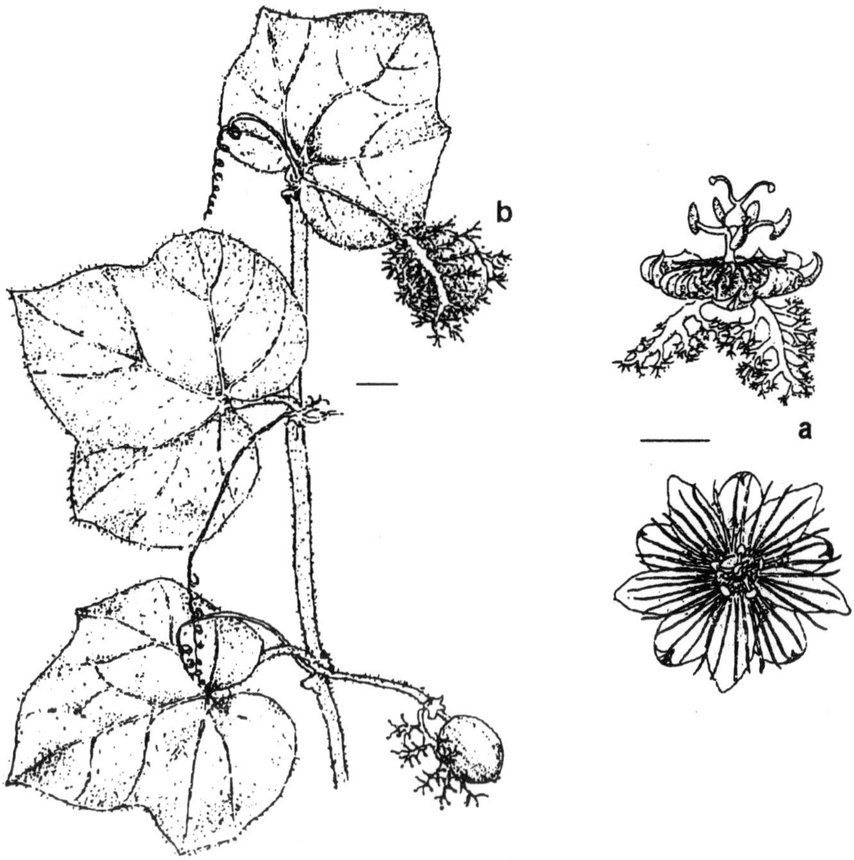

PLUMBAGINACEAE

Aegialitis annulata R.Br. [not illustrated]

(Fl. Mal. **4**: 108; Frodin et al: 1)

Dwarf tree to 1 m high, base of trunk swollen. Twigs with conspicuous annular leaf-scars. Leaves simple, alternate; petiole in-rolled and sheathing stem at base, with slime in axil, longer than blade; blade broadly ovate to ± trullate, 4-8 x 2.5-7 cm glabrous, ± fleshy, with numerous salt glands. Flowers in branched inflorescence; corolla 5-lobed, white. Fruit a pencil-like but slightly curved capsule, 5 x 0.3 cm, 5-sided, dull crimson when mature.

Mangrove, E and W sides.

Lesser Sunda Islands, Moluccas, New Guinea and N Australia.

HCFH 613/13144; Naoni s.n. 15/10/81.

Plumbago zeylanica L.
(Fl. Mal. **4**: 109)

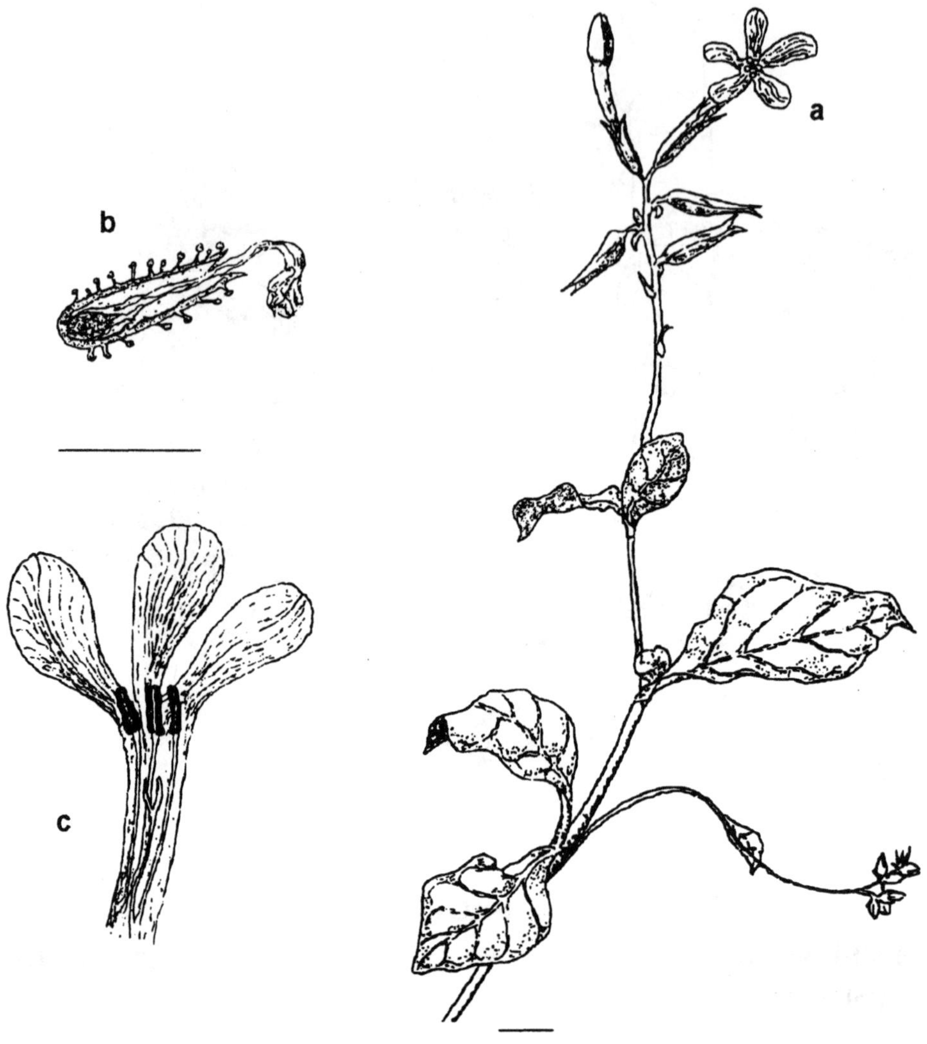

Scrambling herb. Leaves simple, alternate, pale dull green, glabrous; margin slightly wavy; petiole clasping stem. Flowers in terminal spike (a). Calyx green, persistent, 5-ribbed, bearing sticky stalked glands (withered flower in section, b); corolla tubular with 5 lobes, white; stamens 5, blue-purple, in mouth of corolla tube (in section, c). Fruit a 5-angled capsule.

Uncommon in scrub understorey, N end and W side.

Old World tropics, to Hawaii; possibly indigenous only in SE Asia.

HCFH 1970/13742; Huxley s.n. 15/6/75.

PORTULACACEAE

Portulaca pilosa L. [not illustrated]

Succulent, prostrate herb with much branched stems and tufts of white hair at nodes. Leaves simple, spirally arranged, succulent, cylindrical, c. 1 cm long, glabrous. Flowers solitary, terminal, ± sessile; calyx a persistent papery cup; petals 5, yellow. Fruit a dry cone-shaped capsule.

Station area. Widespread in tropics.

Frodin et al. UPNG 5234, 6650.

RHAMNACEAE

Colubrina asiatica (L.) Brongn.

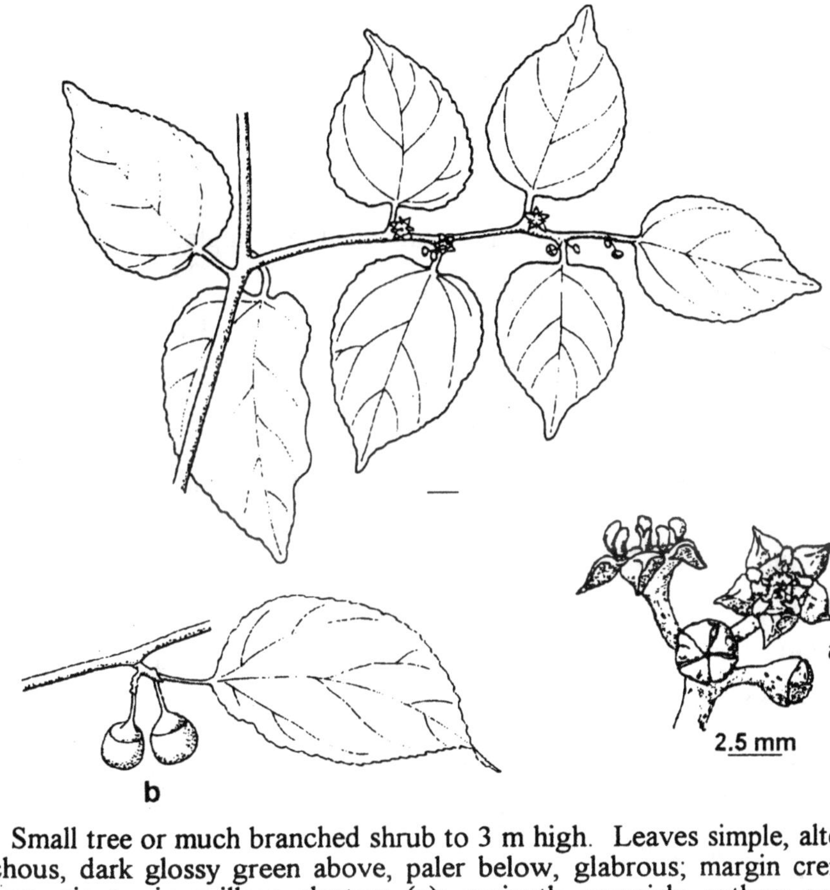

Small tree or much branched shrub to 3 m high. Leaves simple, alternate, distichous, dark glossy green above, paler below, glabrous; margin crenulate. Flowers minute, in axillary clusters (a); perianth greenish; anthers and disc yellow. Fruit (b) green, becoming brown, dry, containing 3 black seeds.

Strand, behind mangroves, E side.

E Africa, Indian Ocean, S Asia, Malesia, Pacific; Caribbean (introduced?).

Frodin UPNG 3189; HCFH 1732/13664; Huxley s.n. 15/6/75; Lawong s.n. 24/8/88.

RHIZOPHORACEAE

Bruguiera gymnorrhiza (L.) Lam. [not illustrated]

(Fl. Mal. **5**: 461; Frodin et al: 30)

Tree to c. 10 m tall. Twigs bearing annular scars. Leaves in upward-pointing clusters. Blade to c. 15 x 6 cm, coriaceous, glabrous, dark green somewhat shiny above, paler below; margin entire, somewhat revolute; petioles reddish. Stipules large, lanceolate, sheathing apical bud. Flowers solitary in axils. Calyx bell-shaped, red, with 12-14 long, pointed lobes, persistent. Hypocotyl cigar-shaped, ridged, dark green, c. 11 x 1 cm, pendent.

Occasional in mangrove, E side.

E Africa, Indian Ocean, S & SE Asia, Malesia to Australia, Micronesia, Polynesia.

HCFH 1090/13622.

Ceriops tagal (Perr.) C.B. Rob. var. **tagal** [not illustrated]

(Fl. Mal. **5**: 469; Frodin et al: 40)

Small tree. Twigs with annular scars. Leaves simple, opposite, elliptical, rounded at apex, blade to c. 7 x 3 cm, glabrous, coriaceous, green to yellow-green, shiny above; margin entire. Inflorescences pedunculate, condensed cymes. Calyx 5-lobed, persistent, green; petals 5, with 3 appendages at apex, white turning brown. Hypocotyl pendent, angled, thickening towards pointed apex, dark green, c. 12 cm long.

Mangrove, occasional on E side.

E Africa, Indian Ocean, S Asia, Malesia, Micronesia, Australia, to New Caledonia.

Frodin UPNG 5237; HCFH 1091/13623.

Rhizophora lamarckii Montr. [not illustrated]

(Fl. Mal **8**: 550; Frodin et al: 54)

Tree. Twigs with annular scars. Leaves simple, opposite, elliptical, glabrous, coriaceous, blade up to 11 x 6 cm, mucronate at apex. Inflorescence on a short peduncle, arising within terminal cluster of leaves; flowers 2-4. Hypocotyls rarely formed.

Mangrove. Rare.

Scattered distribution; known from Sri Lanka, Lesser Sunda Islands, Papuasia, NE Australia and New Calendonia.

Daur s.n. 30/11/82.

Rhizophora stylosa Griff. [not illustrated]

(Fl. Mal. **5**: 456; Frodin et al: 58)

Tree. Twigs with annular scars. Leaves simple, opposite, coriaceous, glabrous, broadly elliptical, obtuse at apex. Inflorescence much branched, peduncle longer than petiole, arising within terminal cluster of leaves, 4 to 16-flowered. Calyx 4-lobed, persistent; petals 4, fleshy, white, hairy, caducous. Hypocotyl pendent, to 40 cm long.

Abundant mangrove on E and W sides.

SE Asia, Taiwan, Malesia, NE Australia, Micronesia, Fiji, New Caledonia.

HCFH 933/13464; Powell et al. UPNG 2395.

Rhizophora apiculata Bl. is also recorded from Motupore by Frodin (no date) but there are no collections from the island at UPNG.

RUBIACEAE

Hedyotis corymbosa (L.) Lam. [not illustrated]

(Henty & Pritchard: 147)

Slender annual herb, weakly ascending, to c. 10 cm. Leaves simple, opposite, lanceolate, c. 1.5 cm long, glabrous; stipules papery, fused to leaf base. Flowers in small, lax, axillary cymes; calyx with 4 tiny teeth; corolla white. Fruit a cup-shaped, papery capsule.

Station area.

Pantropical weed, probably native of Old World.

Frodin et al. UPNG 5229, 6654.

Spermacoce assurgens Ruiz & Pav. [not illustrated]

(Henty & Pritchard: 149)

Small herb to 20 cm. Stems wiry, ribbed, with lines of white hairs, otherwise glabrous. Leaves simple, opposite, narrowly elliptical, to 3 x 0.9 cm, glabrous; petiole short; margin entire; stipules papery, fused to petioles, fringed. Flowers in dense axillary clusters, very small; corolla pinkish.

Station area.

Pantropical weed, probably native to tropical America.

HCFH 1775/13707.

Aidia cochinchinensis Lour.

Randia cochinchinensis (Lour.) Merr.

Small bushy tree to 5+ m high. Leaves simple, opposite, lanceolate or narrowly elliptical, up to 13.5 x 4.5 cm, apex acute, base cuneate, glabrous, mid- to dark glossy green above, paler below; margin entire. Flowers in condensed, repeatedly branching cymes, several arising together. Calyx cup-shaped; corolla tubular, salviform, 5-lobed, cream; anthers large, exserted. Fruit a smooth, ovoid berry with persistent collar at apex, green, ripening red (a).

Rare. A single tree at edge of scrub near station.

China, Taiwan, Indo-China, Malesia and Solomon Islands.

Frodin et al. UPNG 2913 (1977 & 1978); JM 92/4.

Canthium suborbiculare (C.T. White) Merr. & Perry

Small tree or shrub, to 4 m high, much branched. Leaves simple, opposite, ± circular, coriaceous, glabrous, mid- or yellow-green, shiny above, paler and dull below; venation paler, not very distinct; margin entire. Inflorescence many-flowered, axillary, pedunculate, flat-topped (a). Flowers fragrant; calyx/receptacle green; corolla 4-lobed, white; style long-protruding (b), white;. Fruit a spherical berry (c), green ripening black, ± fleshy, containing 2 seeds (transverse section, d).

Common in savanna and scrub.

New Guinea, endemic to Port Moresby region.

Elias MME 15; Frodin UPNG 807, 4245, 5211; HCFH 920/13451.

Guettarda speciosa L.

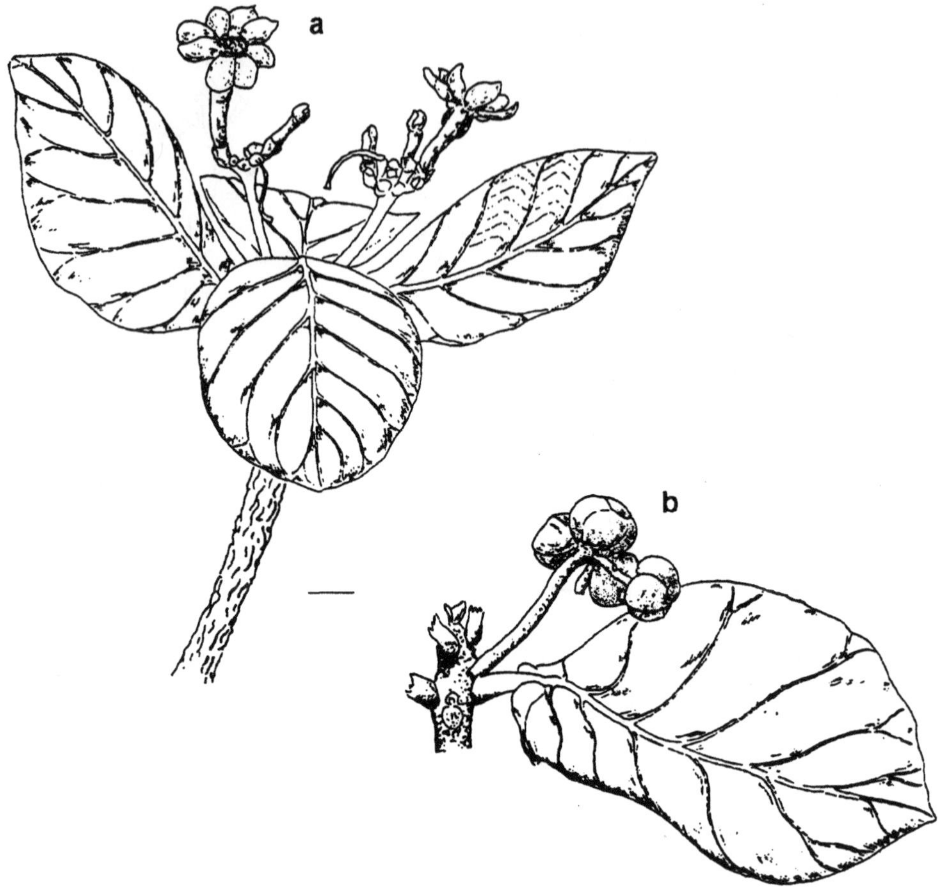

Small or medium tree, to c. 8 m. Stems brown, corky. Leaves simple; opposite, large, broadly obovate to almost circular, cordate at base, ± rounded at apex, light or yellow-green, glossy, glabrous above, paler and sparsely hairy below; main vein thick; margin entire. Stipules small, leaf-like, caducous. Flowers c. 15, in dense pedunculate axillary cluster, only 1 per cluster open at once (a); fragrant. Buds creamy yellow, velutinous. Corolla tubular, white, c. 6-8 lobed. Fruit oblate, mottled pale green, woody, smooth (b).

Strand, especially W side and NW corner; common.

Tropical Asia to SE Polynesia.

Frodin UPNG 3164, 4253; HCFH 619/13150, 1060/13592.

Morinda citrifolia L.

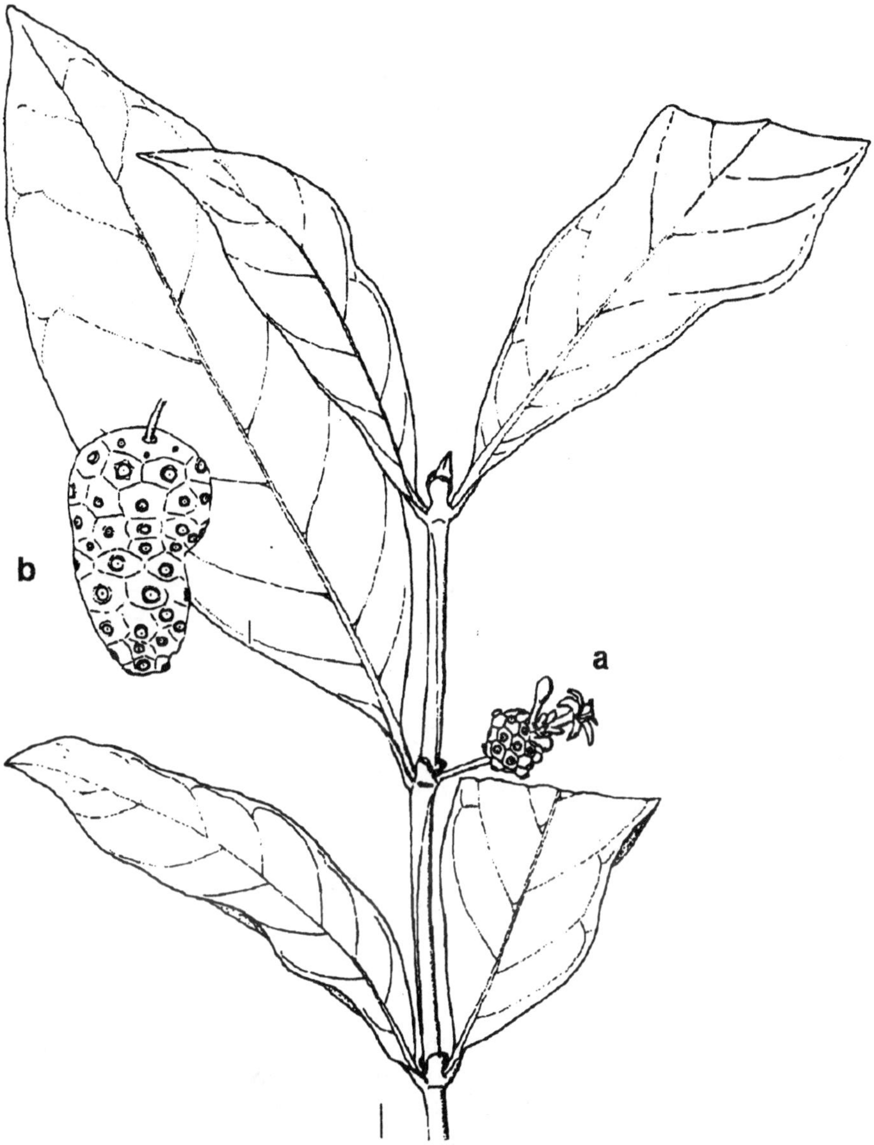

Shrub to 3 m high. Stem 4-sided, glabrous, green. Leaves simple, opposite and decussate, glabrous, green; venation cream, prominent below; margin entire, somewhat undulate; stipules large. Inflorescence pedunculate, axillary, swollen, green (a). Corolla white, tubular, 5-lobed; stigma large, bifid, greenish; faint sweet smell. 'Fruit' compound, green turning opaque, creamy white at maturity, globose, with numerous round brown scars, soft, pulpy, foul smelling (b).

Strand, S end and by the Old Hostel.

Malesia, New Guinea, Australia, some Pacific islands.

Bigilale BH 3; HCFH 2023/13775.

Psychotria bracteosa Valeton

P. lolokiensis Moore

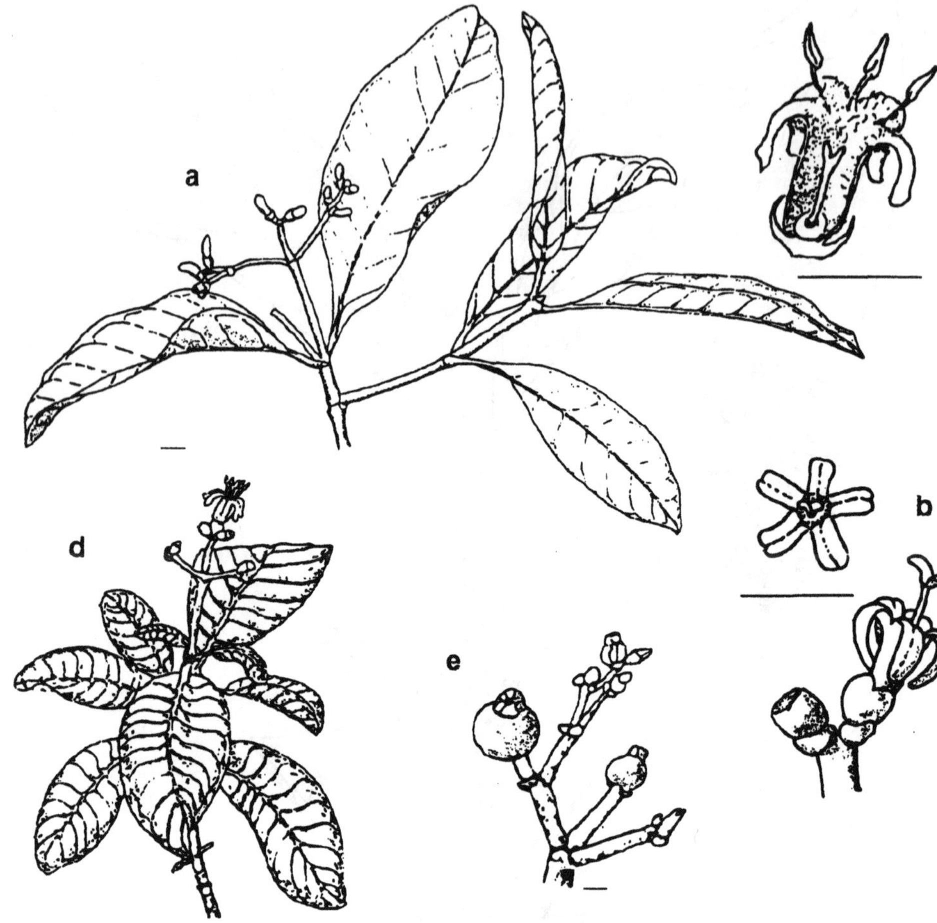

Shrub or small tree, 2-6 m high. Young stems green, old ones pale brown. Leaves simple, opposite and decussate, sometimes a large pair and a small pair alternating; blade elliptical, glossy, mid yellow-green above, greenish white below, glabrous; margin entire. Inflorescence terminal, whitish (a). Flowers of two sorts, on different plants, both bisexual; in both, corolla white, tubular with recurved lobes and the tube hairy inside in upper half and mouth. In "pin" flowers, corolla lobes and stamens usually 5, the stamens enclosed with the corolla tube, the style long, exserted (b). In "thrum" flowers, corolla lobes and stamens usually 6, stamens exserted, style short and enclosed within the corolla tube (half flower c; d). Fruit globose (e), ripening dark crimson with a prominent pink collar, fleshy, containing 2 seeds.

Occasional near shoreline, at edge of savanna or in scrub.

New Guinea.

Frodin UPNG 3193; HCFH 1068/13600, 2005/13757; JM 94/100, 94/120.

Timonius timon (Spreng.) Merr.

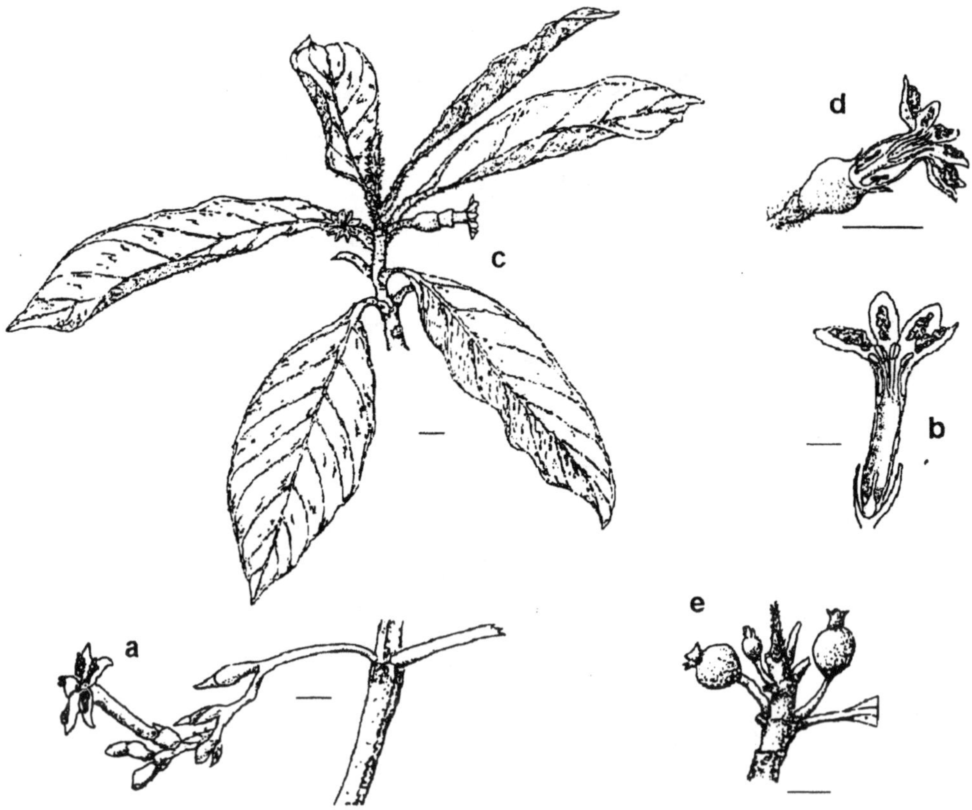

Small, spreading, unisexual shrub or tree, 2-6 m high; female trees rather more robust than males. Young stems softly hairy. Leaves simple, opposite, at twig apices, elliptical, tapering at both ends, pale yellow-green or dull green above, paler below, usually softly hairy on under side, especially on veins, glabrescent; margin entire. Dioecious. Male flowers in cymose clusters (a); calyx 5-toothed, hairy; corolla white, tubular, salviform, 5-lobed, each lobe with central 'crest', hairy on outside, scented; anthers exserted (flower in section, b). Female flowers solitary and somewhat larger (c), with 9-11 corolla lobes and sterile stamens deep in tube, stigmas protruding (flower in section, d). Fruit smooth, ovoid, to 2 cm diameter (e), green, ripening pale brown, containing numerous seeds in dry pulp.

Steep grassy slopes, SE corner and W side, 0-40 m ASL.

Timor, Moluccas, N Australia, New Guinea and Solomon Islands.

Frodin UPNG 3171; HCFH 1067/13599; JM 91/6; Powell et al. UPNG 2324.

RUTACEAE

Glycosmis cyanocarpa (Bl.) Spreng.

G. greenei auct. NG, non Elmer

Small treelet, 2-3 m high. Leaves alternate, compound, 3(-5) leaflets; leaflets elliptical, the terminal one usually the largest, dark green above, paler below; petioles green; venation ± indented above; margins entire; many minute oil glands present. Flowers in branched, axillary inflorescences (a), 9 mm diameter (b); petals white. Fruits < 1 cm diameter (c), immature ones green, ripening creamy pink to orange-red, juicy or fleshy, containing 1-3 seeds.

Occasional in scrub on E side, 1-10 m ASL.

India to Malaya, Borneo, Philippines, Java, Aru, New Guinea.

Frodin UPNG 3175; HCFH 1748/13680; JM 95/136.

Micromelum minutum (G. Forst.) Wight & Arn.

Shrub or small tree, 1-4 m tall. Young stems green with brown blotches, shortly pubescent. Leaves alternate, pinnate; leaflets 7-15, alternate, distinctly asymmetrical, dark green above, paler below, with numerous oil dots, softly hairy with short scattered hairs on both surfaces; venation not distinct above, more prominent below; margin ± crenate; petiole scurfy. Inflorescence flat-topped, branched, shortly pubescent (a). Tepals pale cream-green; filaments white; strong musty smell. Fruit globose-elliptical, c. 0.5 cm diameter (b); exocarp green, ripening orange-red with oil vesicles; single seed embedded in fleshy mesocarp.

Occasional in scrub on E side, and strand line at scrub edge.

India, China, Malesia, N Australia, Melanesia to Polynesia.

Frodin et al. UPNG 2926; HCFH 606/13137, 1752/13684, 1761/13693; Kuduk & Naoni UPNG 6118.

Murraya paniculata (L.) Jack var. **ovatifoliolata** Engl.

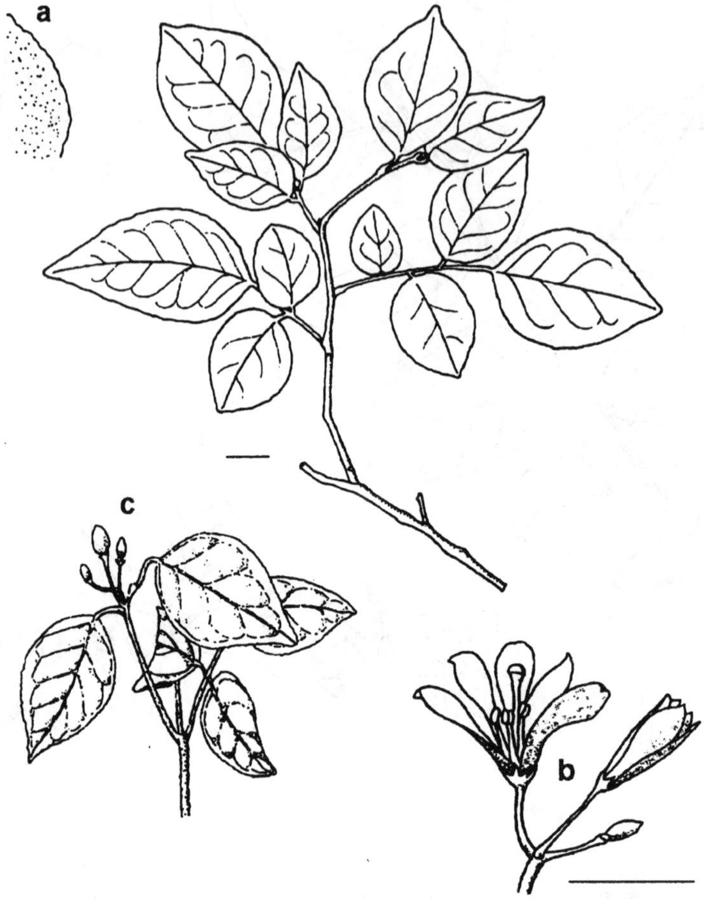

Straggly treelet, 2 m high. Twigs pale whitish-brown. Leaves alternate, pinnate; leaflets 3-8, bright ± shiny green above, dull yellow-green below, gland dotted, ± glabrous; venation obscure; margin crenulate (a); petiole green. Flowers white, fragrant; petals elongate, free (b). Fruit (c) ovoid, 1-1.5 x 0.7 cm; exocarp red with oil vesicles (as in citrus fruits).

Common in scrub at base of nature trail on E side, 1 m ASL.

M. paniculata is widespread but var. *ovatifoliolata* is known only from N Australia and New Guinea.

HCFH 2027/13779; JM 95/137.

Exocarpus latifolius R.Br.

Small to medium-sized tree. Young stems dull green. Leaves simple, alternate, mid- to dark green on both surfaces, paler when young, semi-glossy, rather stiff and thick, glabrous, very variable in size, dotted with numerous tiny white marks; venation parallel, rather obscure; margin entire. Flowers in short axillary spikes (a), minute (b); corolla yellowish, turning red-brown. Fruit in two parts (c): basal fleshy part (swollen pedicel) pinkish-red, with minute white dots; apical part a drupe, hard, yellow-green at maturity. A root parasite.

Common in scrub on E and W sides, and on ridge top.

E Malesia and N Australia.

Elias MME 23; Frodin UPNG 849; HCFH 649/13180; Lawong s.n. 24/8/88.

Alectryon repandodentatus Radlk.

(Fl. Mal. **11**: 457)

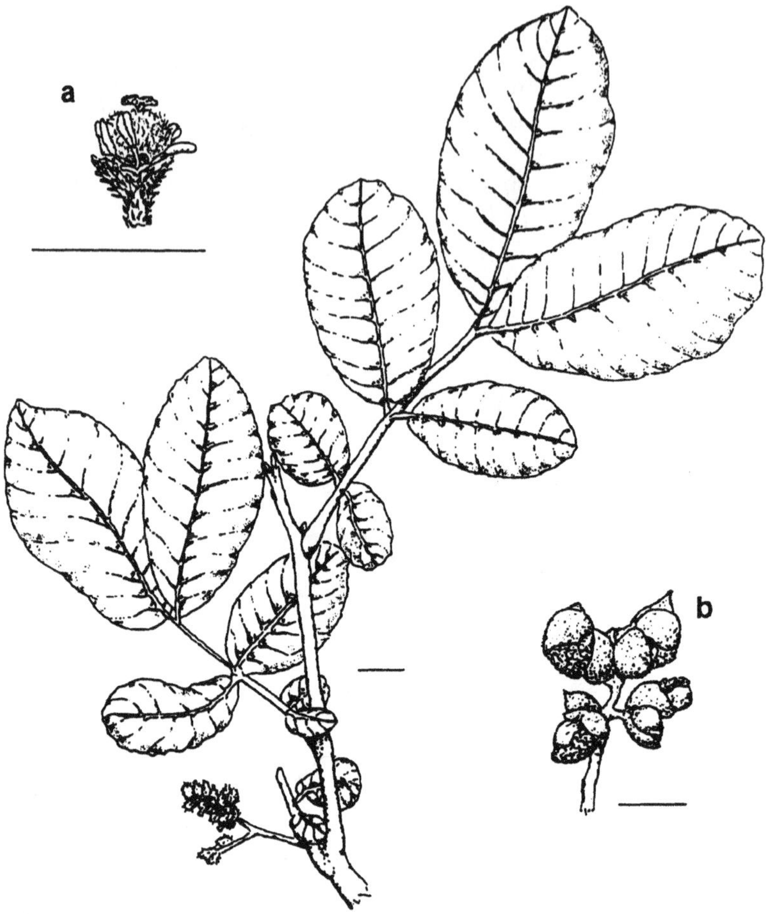

Small tree to 5+ m high. Stems brown. Leaves alternate, pinnate; leaflets opposite (or subopposite), usually in 3 pairs, mid- to dark green on both sides, minutely rusty-pubescent below, glabrescent above; margins with irregular small teeth, revolute; petiole with rusty pubescence. Flowers in branched axillary spikes, minute (a), often unisexual. Pedicel, perianth and ovary all very hairy, pale brown; stamens 7, each with a nectary at base. Fruit green, bilobed, hairy, splitting transversely to reveal one seed in each lobe (b). Seeds black, partly covered by a red aril.

Common in scrub, at scrub edge and as pioneer in savanna.

Endemic to SE New Guinea around Port Moresby and Murray Island in Torres Strait.

Frodin UPNG 806, 4249; Hehonah 1; HCFH 615/13146, 620/13151, 652/13183, 927/13458; Huxley s.n. 15/6/75; Maniwavie TMM 29; Powell et al. UPNG 2340, 2343.

Allophylus cobbe (L.) Raeusch. *sens. lat.*

(Fl. Mal. **11**: 459)

Two distinct forms of this species occur on Motupore.

Allophylus cobbe (L.) Raeusch. entity ***'timorensis'*.**

A. timorensis (DC.) Blume = *A. cobbe* var. *marinus* Corner

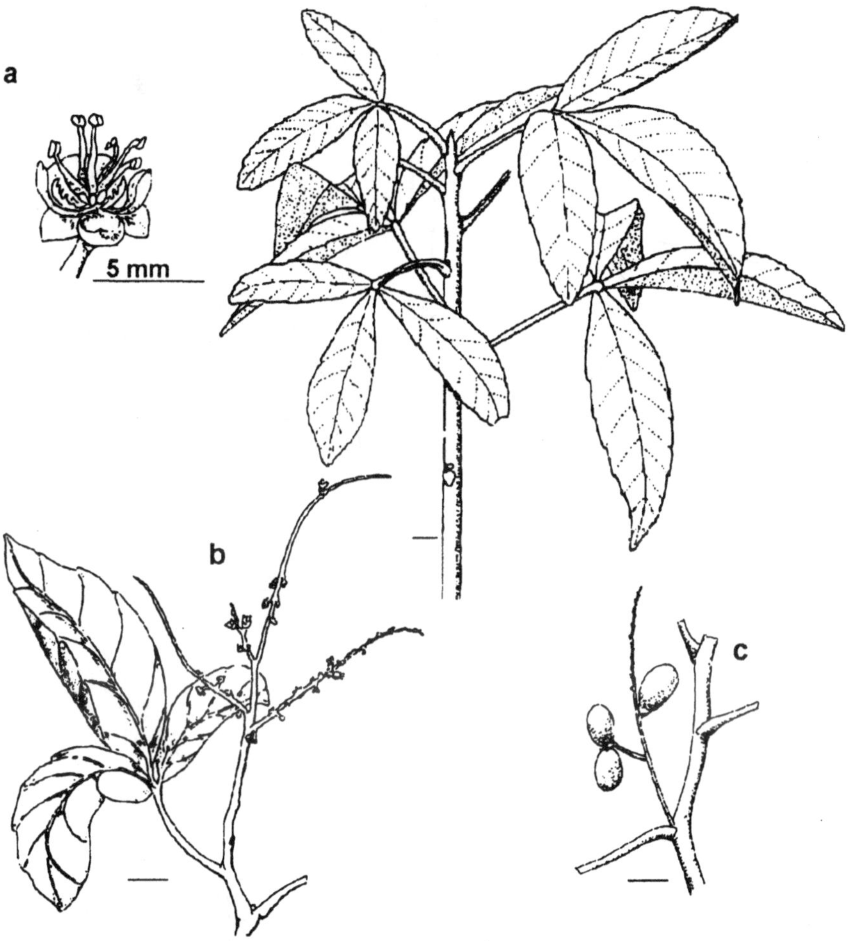

Small shrub c. 2 m high. Twigs pale brown, glabrous. Leaves spirally arranged, trifoliate, yellowish-green above, somewhat shiny, glabrous except occasionally on margins; venation yellow; margins of leaflets crenate; petiole yellowish and grooved. Flowers often unisexual, minute (a), in branched spikes (b); sepals 4, greenish-white; petals 4, white, hairy; nectaries 4, yellow; stamens 8. Fruit oval, yellowish, ripening red, 1 cm long (c).

Occasional in savanna on ridge top, growing in full sun and in scrub on west side.

Frodin UPNG 2914; HCFH 1717/13649, 2013/13765; JM 94/109, 94/126, 94/128.

Allophylus cobbe (L.) Raeusch. entity **'*ternatus*'**.

This differs from the preceeding form as follows:

Liane or sprawling shrub to 4 m high. Leaves up to 2x larger and more widely spaced on stems, dark glossy green above; margin of leaflets entire at base and crenate towards apex (a). Flowers not seen. Fruit (b) succulent, with red pericarp and flesh, containing a single seed.

In scrub on both E and W sides of island.

HCFH 1718/13650; JM 94/127.

Allophyllus cobbe sens. lat. is pantropical.

Cupaniopsis curvidens Radlk.

(Fl. Mal. **11**: 499)

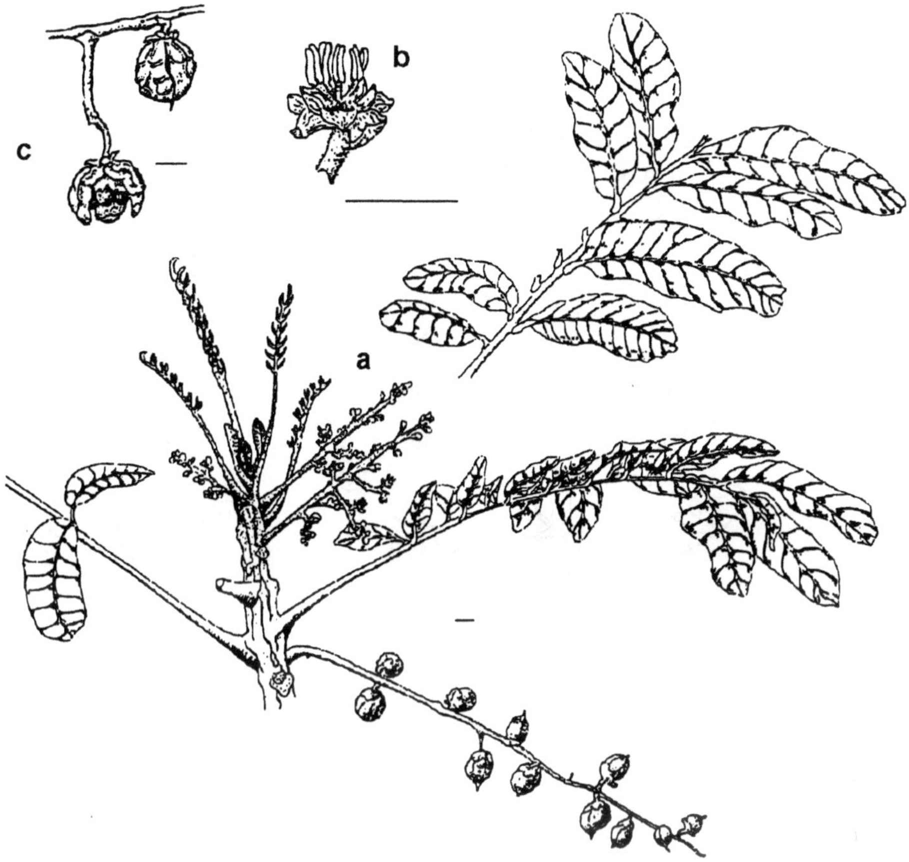

Little branched tree, sometimes multi-stemmed, to 8 m high. Twigs, petioles, peduncles, and young leaves densely hairy. Leaves spirally clustered at ends of thickened twigs giving 'rosette' appearance, pinnate; leaflets c. 7-10 pairs, subopposite, mid yellow-green or grass-green above, paler below, with short brown hairs on underside and on rachis, glabrescent above; margin crenate or shallowly serrate. Inflorescences towards end of twigs, but below youngest leaves, branched towards base (a). Flowers small (b), often unisexual; sepals 5, hairy, brown; petals 4, pinkish-white; stamens 8, conspicuous, filaments hairy. Fruits green, ripening orange-rust brown, c. 2 cm diameter, globose, densely hairy, dehiscent (c); seeds black, partly covered by an orange aril.

Scattered on slopes, E and W sides, margins of scrub patches.

Endemic to New Guinea.

Frodin UPNG 2923, 3183, 5225; HCFH 1742/13674.

Closely resembles *C. napaensis* Adema which is endemic to the Port Moresby area and which also has an orange aril but the leaflet margins on Motupore trees are crenate or dentate, not entire.

Elattostachys tetraporandra Radlk.
(Fl. Mal. **11**: 533)

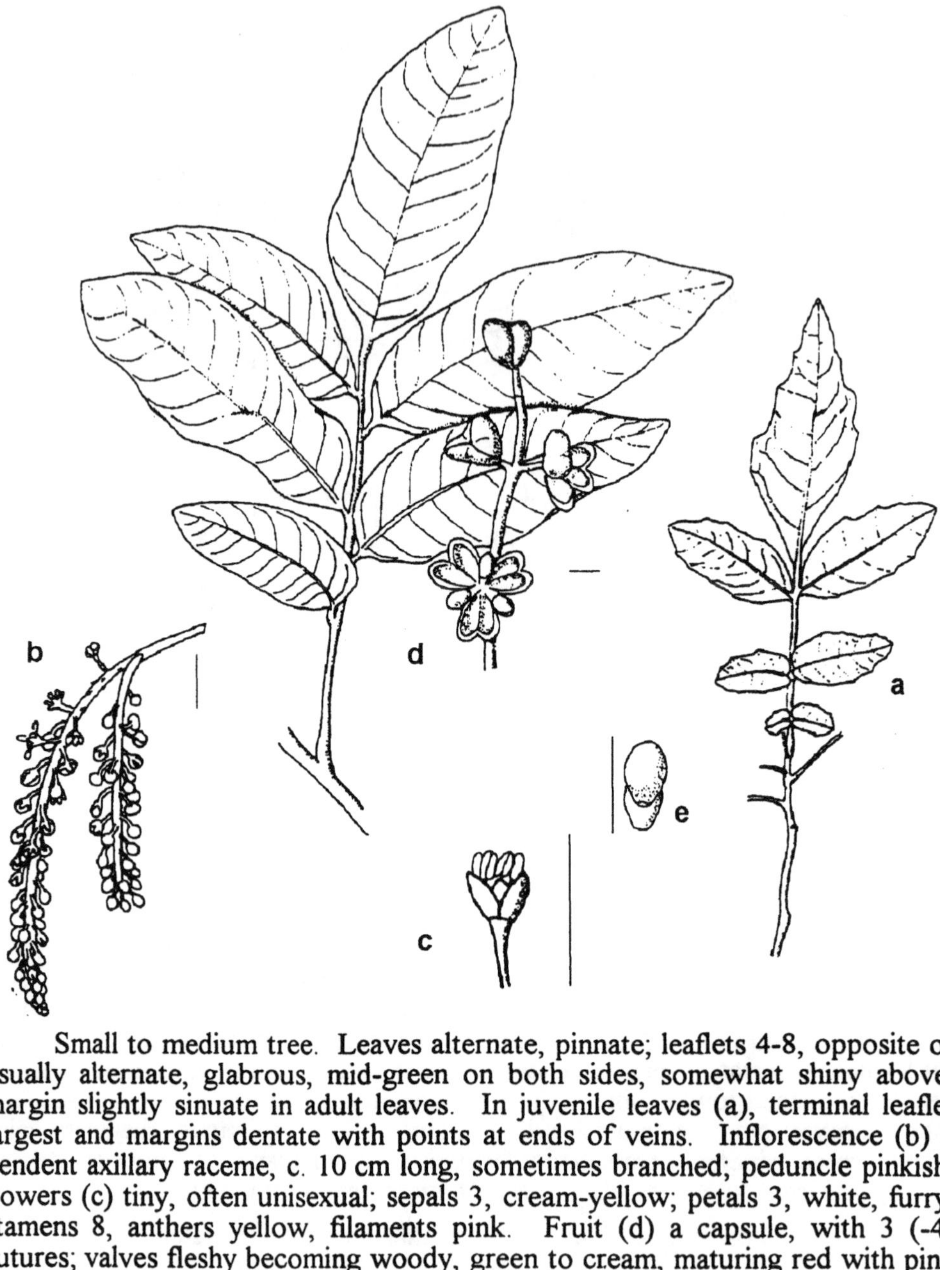

Small to medium tree. Leaves alternate, pinnate; leaflets 4-8, opposite or usually alternate, glabrous, mid-green on both sides, somewhat shiny above; margin slightly sinuate in adult leaves. In juvenile leaves (a), terminal leaflet largest and margins dentate with points at ends of veins. Inflorescence (b) a pendent axillary raceme, c. 10 cm long, sometimes branched; peduncle pinkish; flowers (c) tiny, often unisexual; sepals 3, cream-yellow; petals 3, white, furry; stamens 8, anthers yellow, filaments pink. Fruit (d) a capsule, with 3 (-4) sutures; valves fleshy becoming woody, green to cream, maturing red with pink sutures, inside deep pinkish-red; seed (e) shiny black with small white aril.

Coastal scrub and savanna, common.

New Guinea.

Frodin UPNG 846, 4243; Hehonah 2; HCFH 651/13182, 1061/13593, 1727/13659 (juvenile).

Harpullia leptococca Radlk.

(Fl. Mal. **11**: 608)

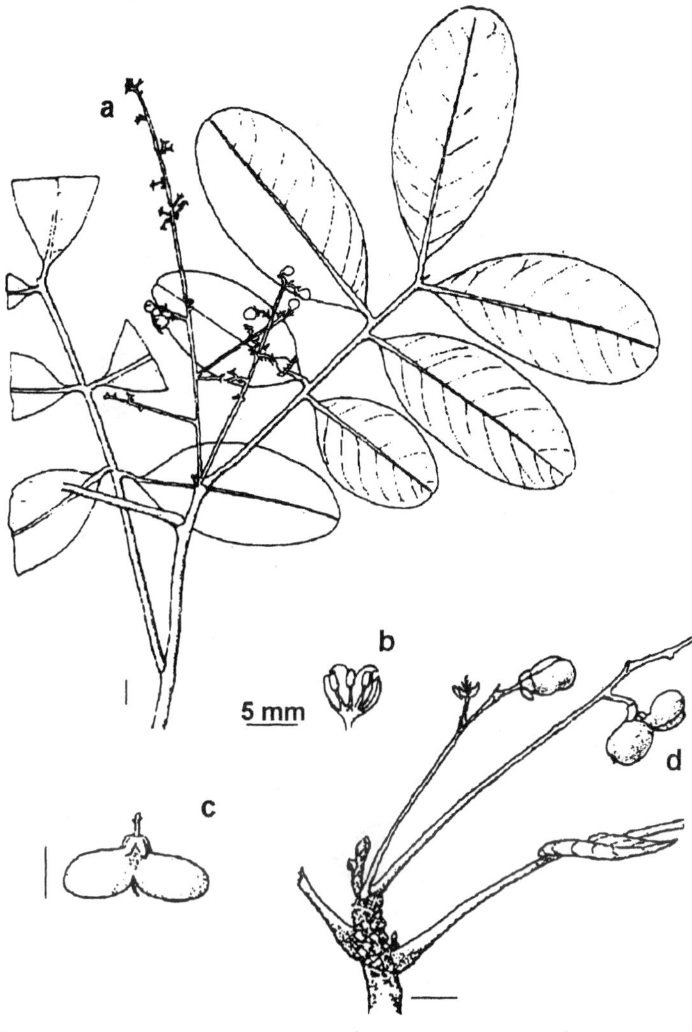

Tree to 8 m high. Leaves spirally arranged, pinnate; leaflets usually in 3 opposite or subopposite pairs, darkish green or yellow-green above, slightly paler below, glabrous, ± coriaceous. Inflorescence an axillary, lax panicle with numerous tiny bracts along the axes (a). Flowers small, dull yellow, often unisexual; calyx lobes 5, velvety; petals 5, white, reflexed at tip; stamens 5; disc between stamens and corolla (section, b). Fruit a bi-lobed, orange capsule, glabrous, inflated with a persistent style (c), dehiscing to reveal 2 black seeds each 3/4 enclosed in a pink-red aril (d).

Common in scrub, at savanna edge, and on strand.

Endemic to Port Moresby region.

Frodin UPNG 3178, 4290; HCFH 1063/13595, 1084/13616, 1739/13671; Powell et al. UPNG 2330.

Mimusops elengi L.

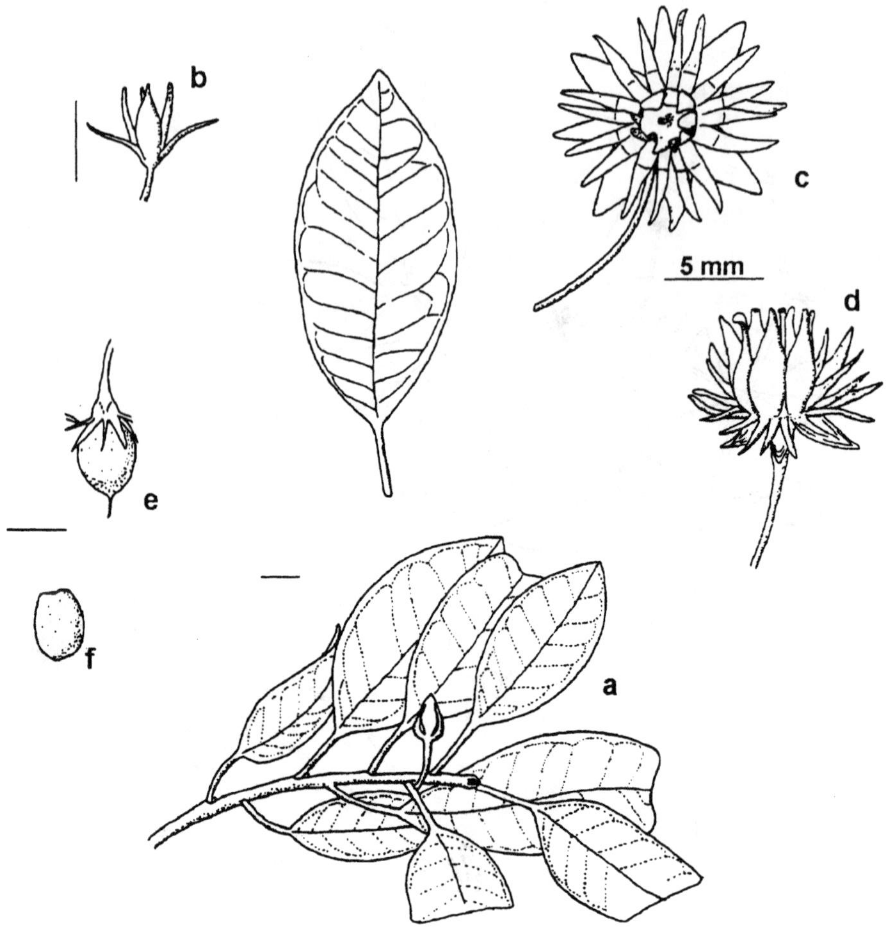

Small or medium-sized bushy tree to 10 m. Sometimes reported as having milky sap. Bark pale reddish brown. Leaves simple, alternate, dark glossy green above, paler below, often conduplicate, glabrous; mid-rib white; margin entire. Young leaves silky. Flowers solitary in axils (a); sepals 8, 4 broad, 4 narrow, brown (b); petals in 2 whorls, inner ones white, ± erect, outer ones white at base and brown at tips, spreading; stamens numerous, brown; style single, yellowish (face view, c; lateral view, d). Fruit a berry with persistent style and calyx (e); immature pericarp green with brown scales, maturing red; flesh yellow, sweet, floury; seeds 1 or 2 with hard, brown testa (f).

Frequent in scrub on E and W sides; also on ridge top.

India to Malesia, N Australia and the Pacific as far E as Hawaii.

Frodin UPNG 2912, 4257, 5223; HCFH 617/13148, 1046/13578, 1724/13656, 1759/13691; Huxley s.n. 15/6/75.

Harrisonia brownii A. Juss.

(Fl. Mal. **6**: 207)

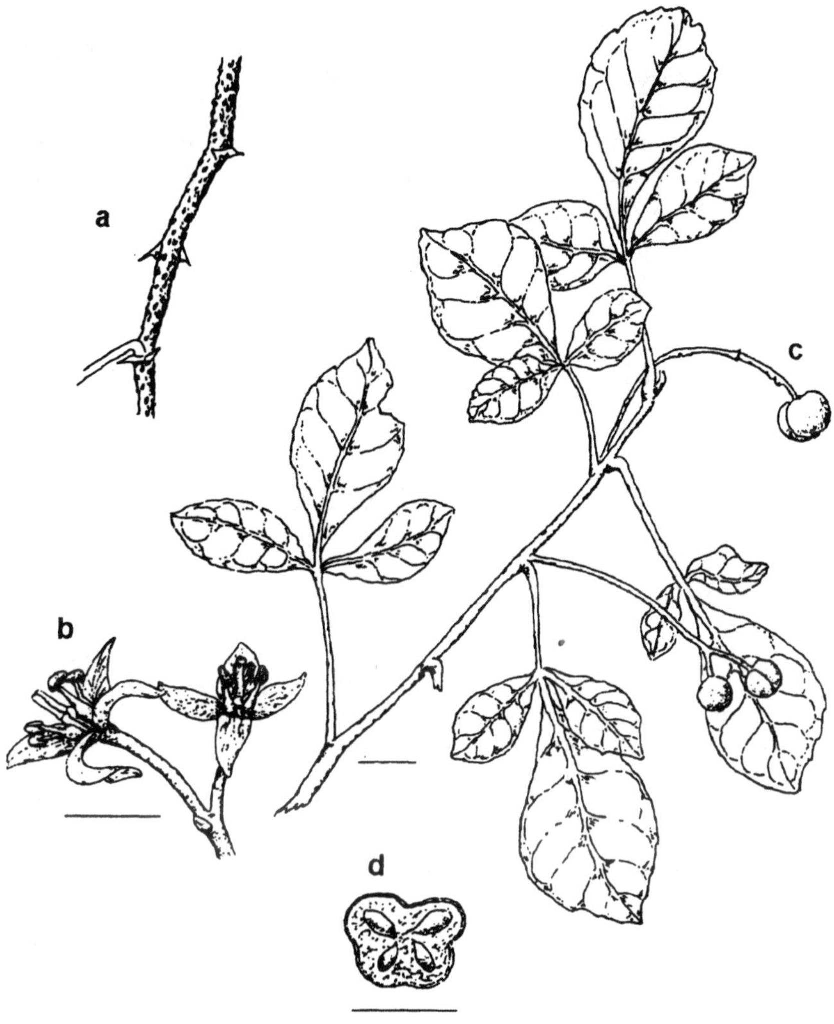

Scandent shrub. Stems brown with numerous lenticels and paired spines at base of older leaves and on leafless twigs (a). Leaves alternate, trifoliate, terminal leaflet larger than laterals, dull mid-green above, slightly hairy on both sides; margin irregularly crenate to serrate towards apex of leaflets. Flowers in axillary cymes, tiny (b); petals 4; stamens 8. Fruit a drupe (c), with shiny black exocarp and purple flesh, slightly 4-lobed, containing 4 seeds (in section, d).

Common in scrub, especially at margins.

Central Malesia to SE New Guinea and N Australia (Gulf of Carpentaria).

Frodin UPNG 5212; HCFH 609/13140, 1062/13594; Leach UPNG 5270; Powell UPNG 2350.

SONNERATIACEAE

Sonneratia alba J. Smith [not illustrated]

(Fl. Mal. **4**: 285; Frodin et al: 64)

Tree. Leaves simple, opposite, broadly elliptical with a blunt apex, attenuate at base, glabrous, thick, coriaceous, dull greyish green; margin entire. Flowers large, solitary or in 3s, nocturnal, pollinated by bats. Calyx lobes 6, triangular, green outside, reddish inside, persistent, spread in fruit; petals white, long and narrow, caducous; stamens white, numerous; style single, far exserted, persistent in fruit. Fruit with a deep cup-like base and sepals radiating from margin, hard, green, containing numerous chambers.

Occasional mangrove, especially on E side.

E Africa, Indian Ocean, SE Asia, Malesia to Micronesia, Melanesia, N Australia.

HCFH 1088/13620.

STERCULIACEAE

Heritiera littoralis Dryand. in Aiton [not illustrated]

(Frodin et al: 44)

Sapling 1-5 m high; adult tree medium to large. Leaves simple, alternate, elliptical, with acute apex and narrowly cordate base, c. 23 x 8.5 cm, dull dark green above, silvery-scaly below; margin entire. Flowers in dense, branched inflorescences, without petals. Fruit hard, brown, keeled, 'boat-shaped'; dispersed by ocean currents.

Rare mangrove on strand line, E side, young plant only.

SE Asia, Malesia, Pacific.

HCFH 1765/13697.

Corchorus aestuans L.

(Henty & Pritchard: 129)

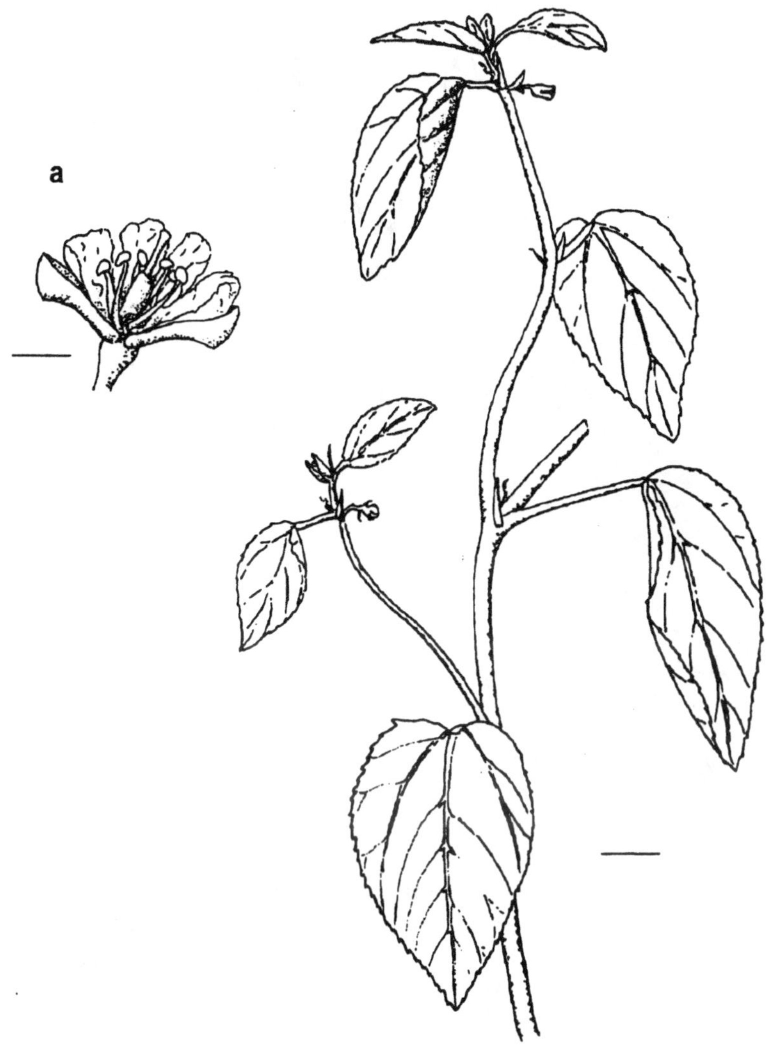

Herb to 0.5 m high, shortly hairy all over. Leaves soft, ovate, serrate. Stipules 2, one lanceolate, one filiform. Flowers in clusters opposite leaves (drawn with one petal removed, a). Sepals 5, green, boat-shaped; petals 5, free, yellow; stamens 10-20. Fruit an elongate capsule.

Uncommon, in shade on ridge top, especially where previously burnt.

Pantropical weed.

JM 94/104.

Grewia acuminata Juss.

a

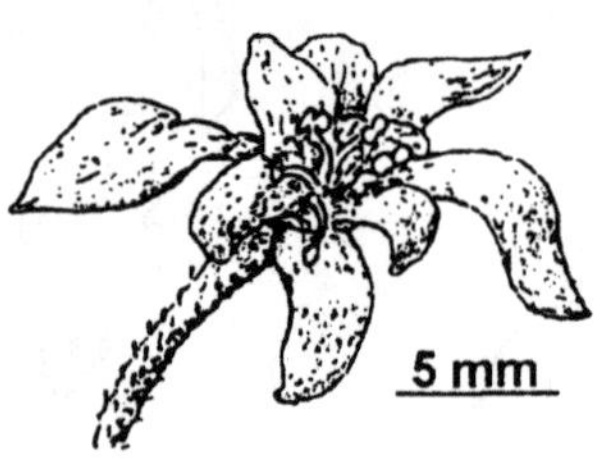

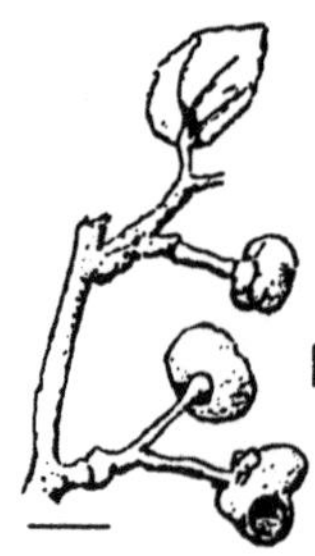

b

Shrub with long trailing or scrambling branches. Twigs red-brown with white lenticels. Leaves simple, alternate, dull or glossy mid-green on both sides, paler below, ovate, length-breadth ratio 1.95-2.3, apex attenuate and base rounded or slightly cordate; stellate hairs especially on lower side at junction of main vein with side veins, otherwise glabrous above and with tiny scattered hairs below; usually 3 large veins arising from junction of blade and petiole; margin shallowly crenulate to minutely serrate. Flowers (a) in few-flowered axillary clusters; pedicels c. 1 cm; calyx lobes 5, 1.5 cm long, densely hairy on outside, dull orange inside; petals 5, c. 0.5 cm long; stamens numerous, c. 1 cm long; ovary and young fruit densely covered with white hairs; stigma 2-lobed, green. Mature fruit (b) brown, hairy, dry, globose, 1.5 cm diameter, 4-seeded.

Coastal scrub, common on E side behind mangroves, and W side.

E Asia and Malesia.

Frodin UPNG 4486, 5218; HCFH 1072/13604, 1763/13695; Kuduk & Naoni UPNG 6124.

Grewia species A

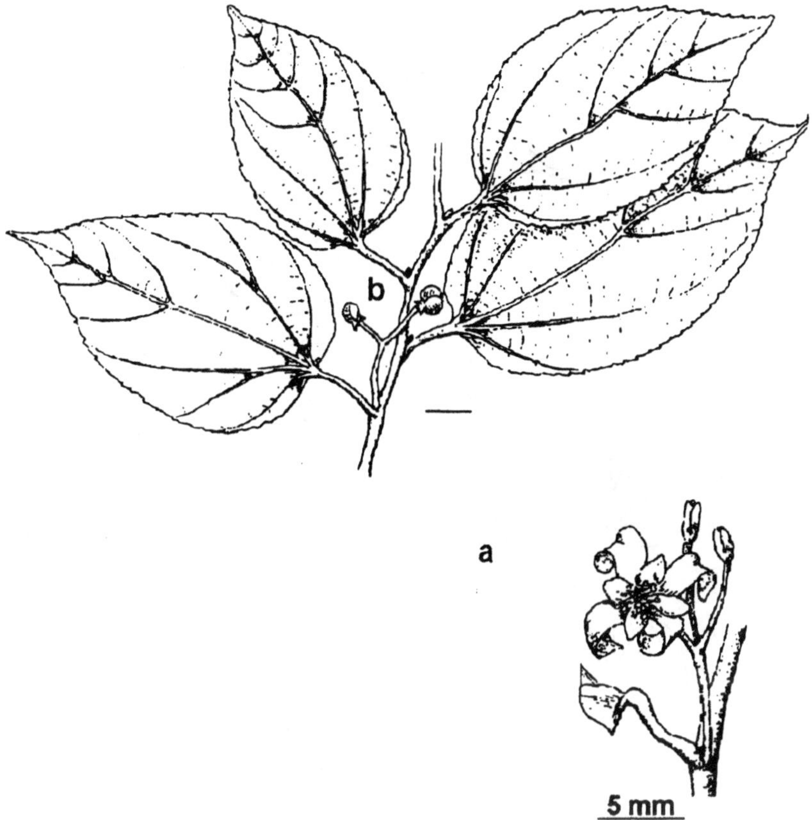

Deciduous tree to 7 m tall. Twigs dull brown, sometimes slightly zig-zag. Leaves simple, alternate, distichous, dull olive-green, cordate, length-breadth ratio 1.56-1.96, apex attenuate and base cordate; stellate hairs on petiole, veins and scattered on blade, especially on lower side; 3 large and usually 2 small veins arising from junction of blade and petiole; margin distinctly crenate. Flowers in few-flowered axillary clusters (a), initially lemon yellow, turning pink and orange with age; pedicels c. 0.7 cm; calyx lobes 0.7 cm, hairy on outer surface, coloured on inner surface; petals c. 0.4 cm, papery when dry; stamens numerous, c. 0.4 cm long. Fruit reddish brown, slightly lobed, c. 0.6 cm diameter (b).

Scrub, E side.

HCFH 705/sn, 1719/13651, 2004/13756.

Resembles *G. breviflora* Benth. or *G. xanthopetala* F. Muell. ex Benth., both of which occur in monsoon forest in northern Australia.

Celtis philippinensis Blanco

(Fl. Mal. **8**: 62)

Shrub or small tree to 10 m high. Twigs green; bark pale. Leaves simple, alternate, distichous, glossy green, or sometimes dark green above, paler and dull below, glabrous, coriaceous, rather variable in size, ovate-elliptical, sometimes asymetrical at base; distinctly 3-veined; margin entire (a) or rarely spinose (b). Flowers in small axillary clusters, small, white. Fruit (c) green, ripening red, smooth, containing a single seed.

Common in scrub.

Africa, Indian Ocean, SE Asia, Malesia, Australia to Solomon Islands.

Frodin UPNG 848, 2924, 2944, 4262; HCFH 1069/13601, 1075/13607; Huxley s.n. 15/6/75; Kuduk & Naoni UPNG 6120.

Trema cannabina Lour.

(Fl. Mal. **8**: 50)

Shrub with several upright stems, 1-2 m high. Leaves distichous, ovate-lanceolate, margin serrulate (a). Inflorescence axillary, 10-12 flowered. Flowers (b) minute, unisexual; tepals greenish white. Fruit (c) a drupe, 3x2 mm, ripening orange (in transverse section, d).

Seen rarely, on west slope, in grass.

SE Asia through Malesia to Fiji, Samoa, Micronesia.

UPNG 12202; JM 95/142.

Avicennia eucalyptifolia Zipp. ex Moldenke. [not illustrated]

A. officinalis L. var *eucalyptifolia* Valet.
A. marina (Forsk.) Vierh. var. *resinifera* (Forst.) Bakh. pp. "*A. resinifera*"

(Frodin et al: 18)

Bushy tree to 3 m. Leaves simple, opposite, narrowly ovate, apex acute, to c. 11 cm long, yellowish-green above, whitish below, glabrous; margin entire. Flowers in branched inflorescences at ends of leafy twigs, small, yellow. Fruits greenish white, smooth-velvety, pointed, c. 1.5 cm long.

Occasional mangrove at N and S ends and W side.

Philippines, Lesser Sunda Is., Moluccas E through New Guinea to the Solomon Islands and Australia.

Frodin UPNG 4261; HCFH 1092/13624; Powell et al. UPNG 2335.

Stachytarpheta jamaicensis (L.) Vahl. [not illustrated]

(Henty & Pritchard: 164)

Woody herb to 1 m tall. Leaves simple, opposite, elliptical with very attenuate base; margin coarsely serrate. Flowers in terminal spikes to 25 cm long, blue, zygomorphic.

Occasional at edge of open ground, Station area.

Pantropical weed, probably of American origin.

HCFH 1099/13631.

Clerodendrum inerme (L.) Gaertn.

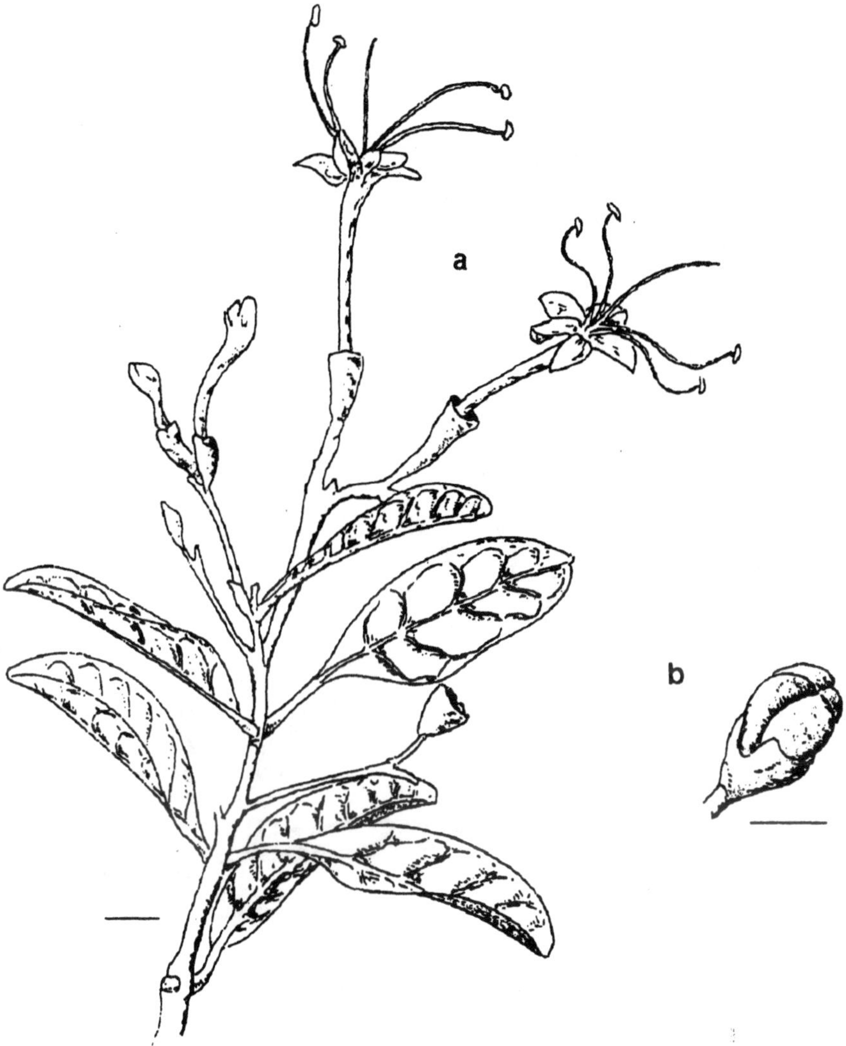

Sprawling shrub to 3 m high. Leaves simple, opposite, yellowish green or light green, somewhat glossy above, paler below, glabrous, gland-dotted, somewhat conduplicate; margin entire. Petioles purple. Flowers in clusters at ends of leafy twigs, showy (a); calyx green, cup-like, persistant; corolla off-white; stamens and style purple. Fruit green at maturity, woody, 4-lobed (b).

Common along strand, just above high water. Also on hill slope at 40 m ASL.

Tropical Asia to W Polynesia and N Australia.

Frodin UPNG 3165; HCFH 1744/13676; Huxley s.n. 15/6/75.

Clerodendrum floribundum R. Br. var. **ovatum** (R. Br.) Domin.

C. populneum Bear & H.J. Lam.

Straggly shrub or small tree to 3 m high. Young stems and petioles red-purple, turning to green then brown. Bark thick, deeply furrowed. Leaves simple, opposite, broadly ovate to cordate, purplish green or olive-green, glabrous, gland-dotted on lower side; margins and veins reddish or yellow; margin ± entire. Flowers in small axillary clusters, white, without fragrance (a); calyx green, with 5 teeth, persistent; corolla white to cream, 5-lobed; stamens 4, white. Fruit black, 4-lobed, in persistent fleshy calyx (b).

Savanna on slope and ridge top.

Australia and New Guinea.

Frodin UPNG 3198; HCFH 1715/13647; Huxley s.n. 15/6/75.

Clerodendrum populneum was placed in synonymy with one of the varieties of *C. floribundum* by A.A. Munir (J. Adelaide Bot. Gard. **11**. 1989).

HCFH 1042/13574 may be a further species of *Clerodendrum*.

Premna dallachyana Benth.

Scrambling shrub to small tree, with ragged crown. Twigs grey-brown with longitudinal cracks and pale lenticels. Leaves simple, opposite, dull or glossy green above when fresh, drying blue-black, glabrous, base sometimes slightly unequal. Flowers small, in *Psychotria*-like inflorescence with decussate branching. Calyx irregularly lobed. Fruit small, green, turning black when ripe (a); locules 2.

Savanna and scrub border.

Philippines, New Guinea and N Australia.

Frodin UPNG 5221; HCFH 1713/13645, 1721/13653; JM 95/138.

Premna serratifolia L.

incl. *P. integrifolia* L. and *P. obtusifolia* R. Br.

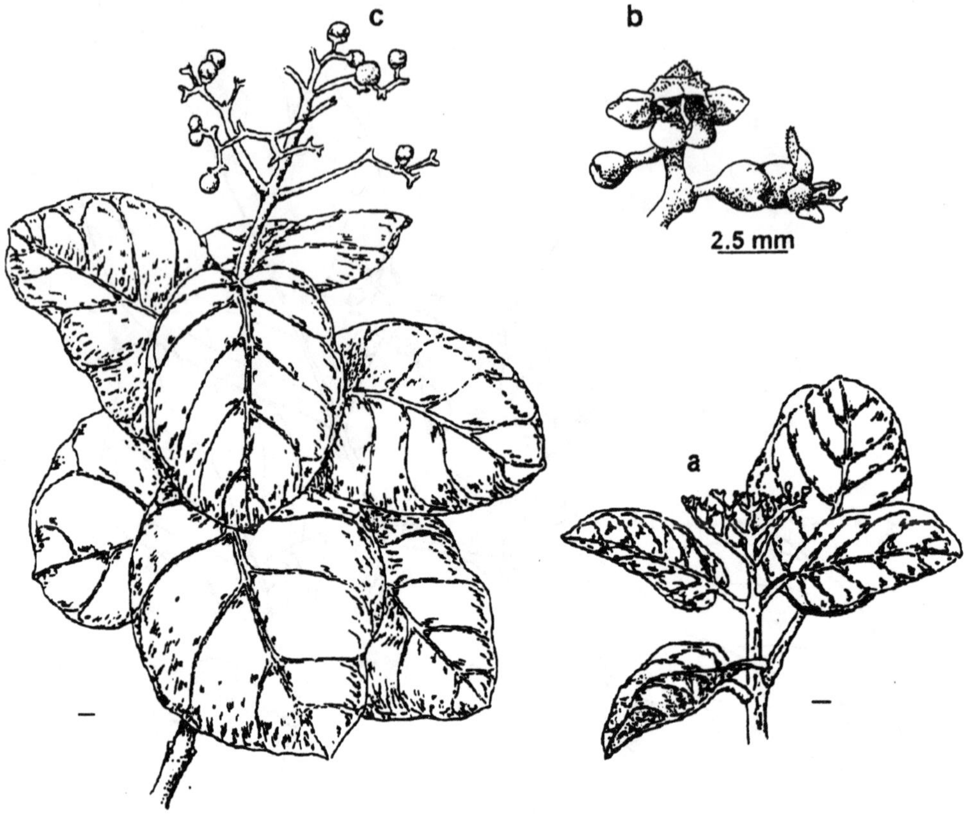

Shrub. Stems pale brown, often with white lenticels. Leaves simple, opposite and decussate, petiolate, broadly elliptical, base rounded or cordate, rather variable in size, yellow-green and shiny above, glabrous; margin entire, somewhat sinuate. Flowers small, in much-branched, flat-topped, terminal inflorescence (a); calyx cup-shaped with 5 shallow lobes; corolla greenish-white; stamens and style exserted (b). Fruit a small, spherical, fleshy, 1-seeded drupe, green ripening black (c).

Common along strand and on slopes, especially W side.

E Africa, S Asia to Pacific and Australia.

HCFH 1065/13597, 1743/13675; Powell s.n. 1972.

Vitex trifolia L.

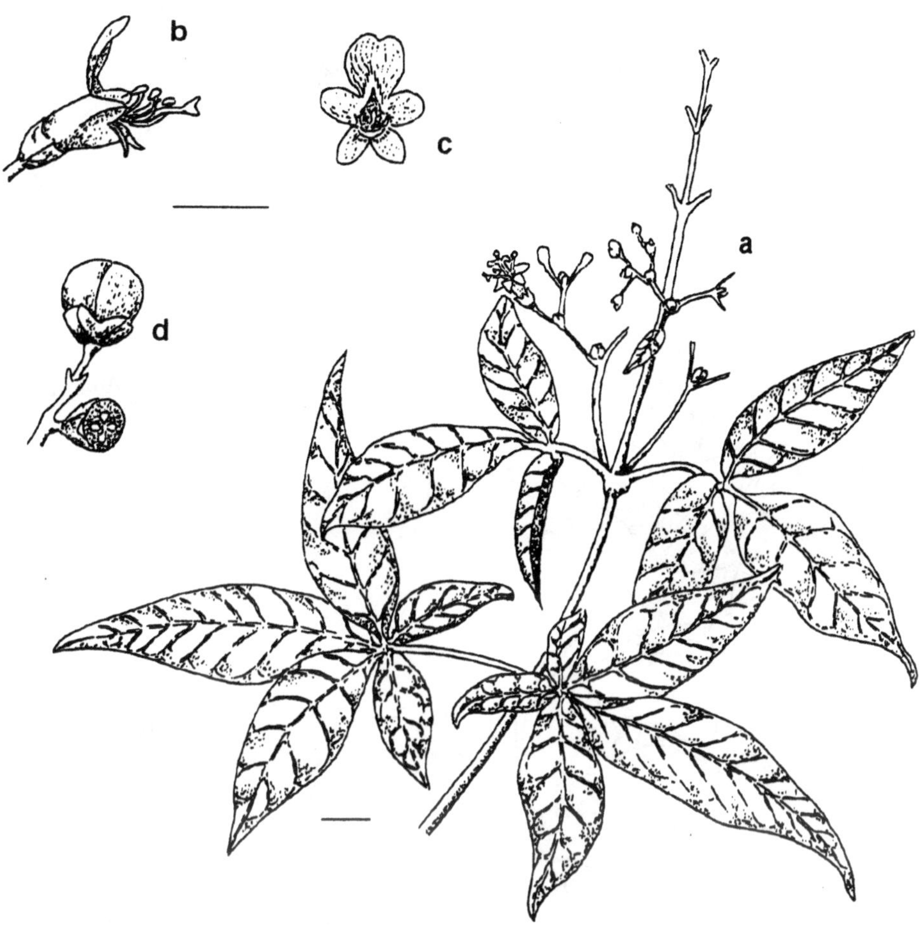

Bushy shrub to 4 m high. Stems square, greenish, minutely hairy, as are petioles. Leaves opposite, palmately compound; leaflets 3-5, light or dark green and glabrous above, whitish below from dense covering of minute hairs; margins entire. Flowers in a branched terminal inflorescence (a), zygomorphic; corolla 5-lobed, blue-violet (lateral view, b; front view, c). Fruit spherical with persistent calyx, containing 4 seeds (d).

At junction of strand and scrub, common on E and W sides.

Africa, Indian Ocean, S Asia, Malesia, Melanesia, N Australia, to Japan, Hawaii and Polynesia.

Frodin UPNG 802; HCFH 616/13147, 1716/13648; Lambinon 87/104; Lawong s.n. 24/8/88.

Ampelocissus acetosa (F. Muell.) Planch.

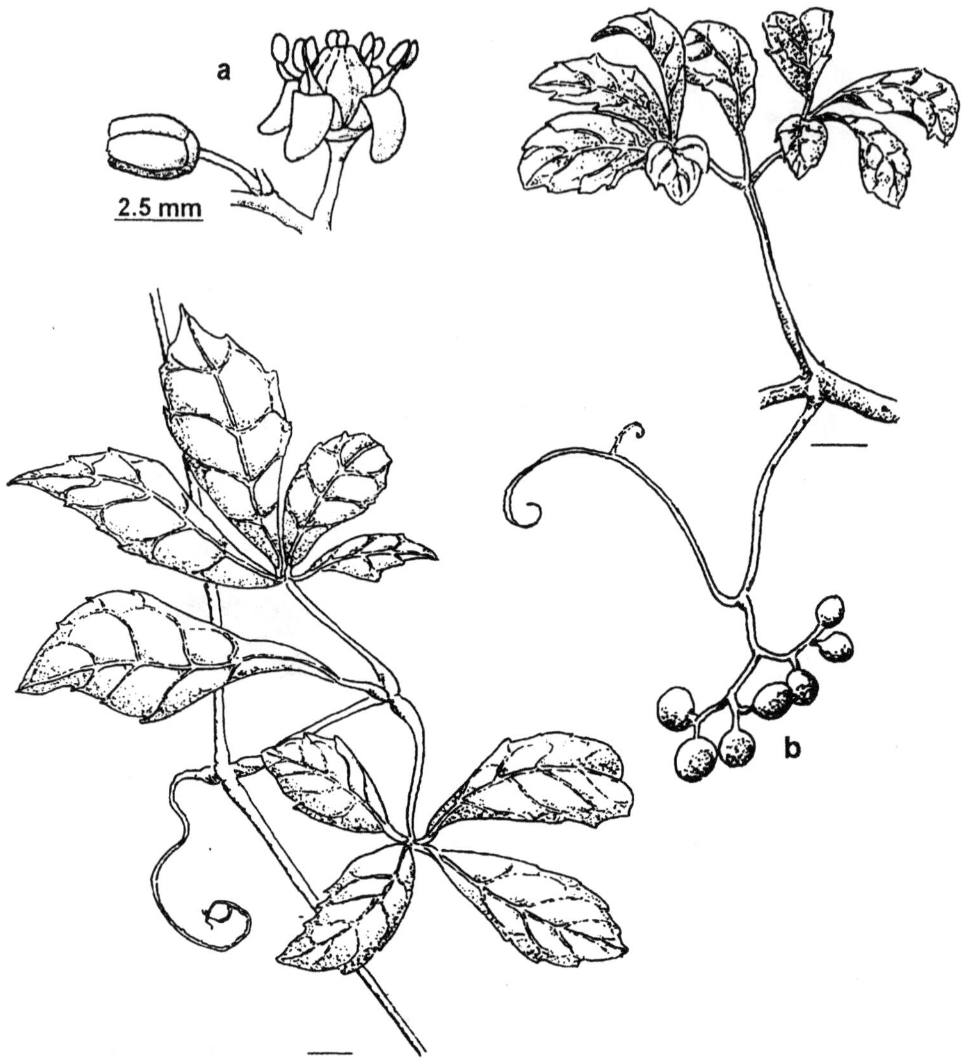

Vine. Leaves pedately compound, alternate; leaflets 5-9; margins with sparse teeth in distal half. Tendril opposite each petiole. Flowers (a) dark purple. Fruit (b) green, turning black, c. 1 cm diameter.

Common at monsoon forest edges or scrambling over shrubs in savanna.

N Australia and New Guinea.

JM 91/7, 95/141.

Cayratia trifolia (L.) Domin

Cissus trifolia (L.) Schum.

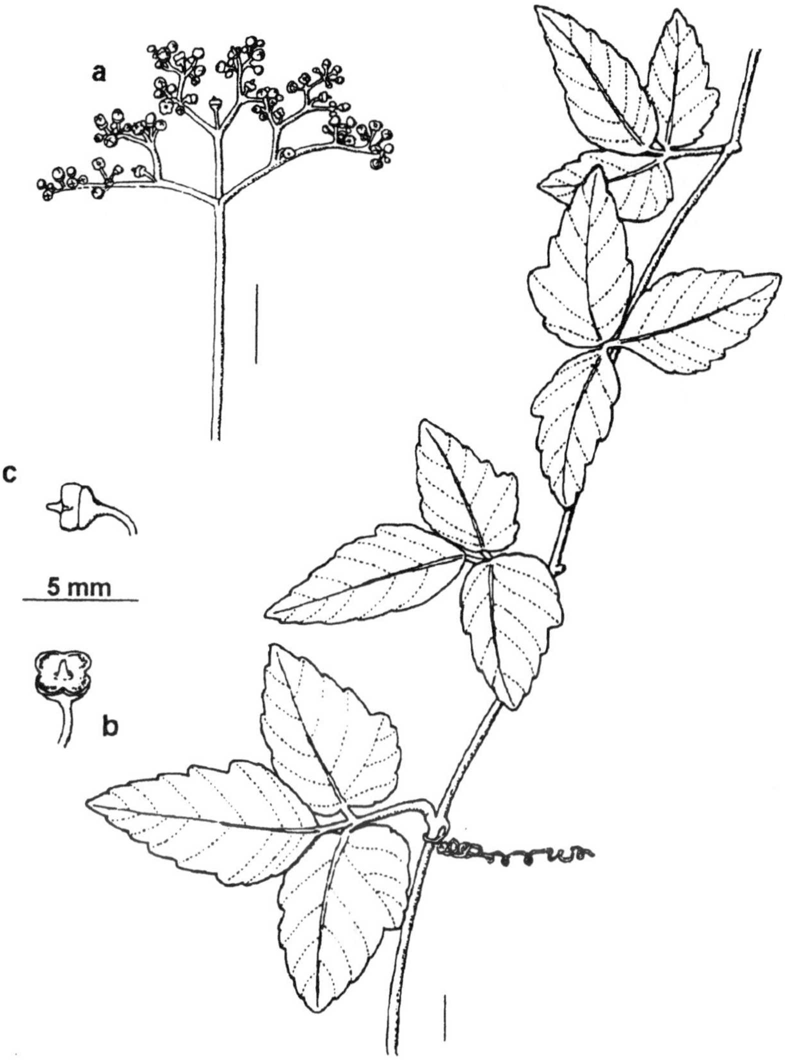

Small tough-stemmed herbaceous creeper. Stems and petioles minutely hairy and ridged. Leaves trifoliate, alternate, a tendril opposite each petiole or sometimes absent; leaflets softly hairy; epidermis white-dotted below. Flowers, bisexual, minute, in much branched terminal umbellate cyme (a), softly hairy. Calyx pale green; corolla pale green to yellowish with 4 free lobes; disk entirely surrounding ovary; style red (front view, b; lateral view, c). Fruit fleshy, c. 0.8 cm diameter, green.

In grass near scrub border at bottom of nature trail on E side.

India to the Pacific.

HCFH 1078/13610.

Cayratia species

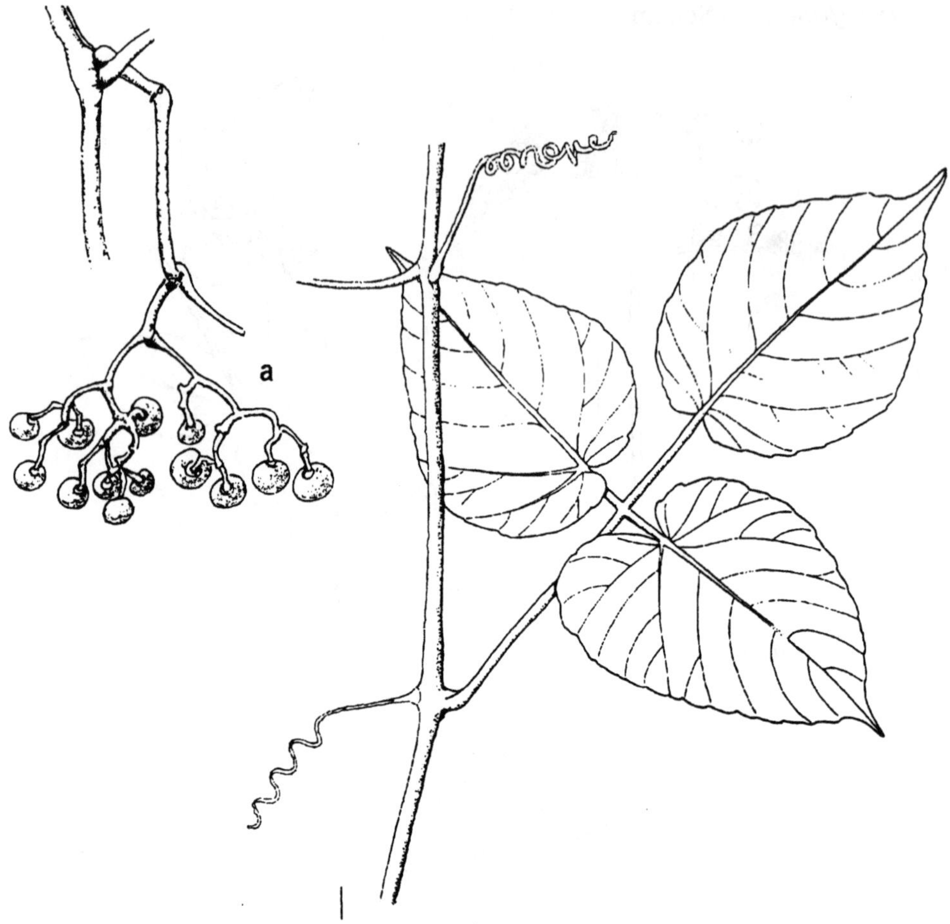

Woody climbers. Stems brown, squarish. Leaves trifoliate, alternate; a tendril opposite each petiole, branched towards apex; fine white hairs on petiole and veins, otherwise glabrous; margin crenate. Flowers minute in much branched, flat-topped, axillary cyme, green, bisexual. Fruits green, turning black, fleshy, 4-seeded (a).

Scrub, E and W side.

A group of rather similar species, including:

Cayratia acris (F. Muell.) Domin (syn. *C. grandifolia* (Warb.) Merr. & Perry), which occurs in tropical Australia, S New Guinea and Solomon Islands. *HCFH 1073/13635*.

Cayratia cardiophylla B. R. Jackes, which occurs in Australia and New Guinea. *JM 91/15*.

Cayratia schumanniana (Gilg) Suesseng., which occurs only in New Guinea. *HCFH 2007/13759*.

Cissus hastata Miq.

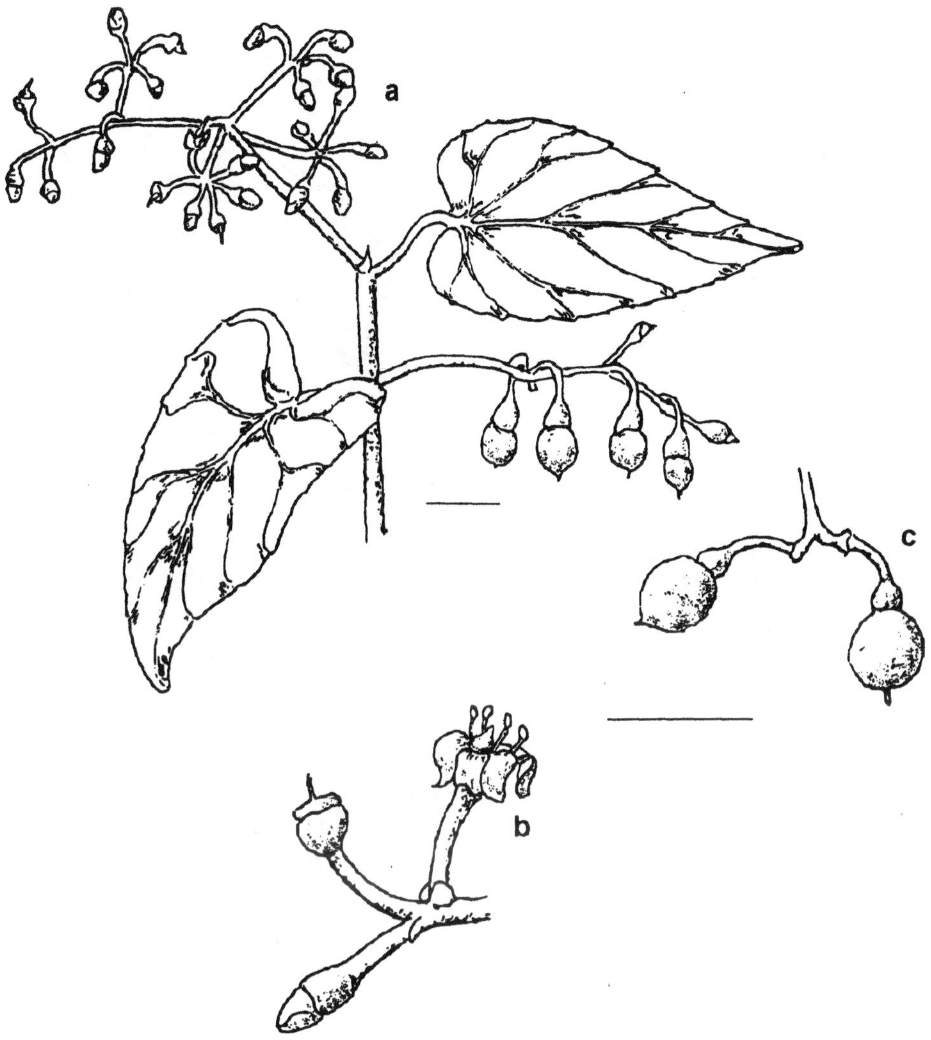

Climber. Stems dark red, 4-angled. Leaves simple, alternate, with a tendril opposite each petiole; blade dull mid-green above, glossy below, glabrous, cordate at base; margin with minute teeth at ends of veins. Inflorescence replaces tendril, branched (a), intense dull red. Tepals reflexed, stamens 4, yellow; style single (b). Immature fruit green, later black, on an enlarged base (c).

Frequent in savanna.

India through to New Guinea and Australia.

HCFH 621/13152, 1055/13587, 1733/13665; Kuduk & Naoni UPNG 6116; Lambinon 87/109.

Cissus repens Lam.

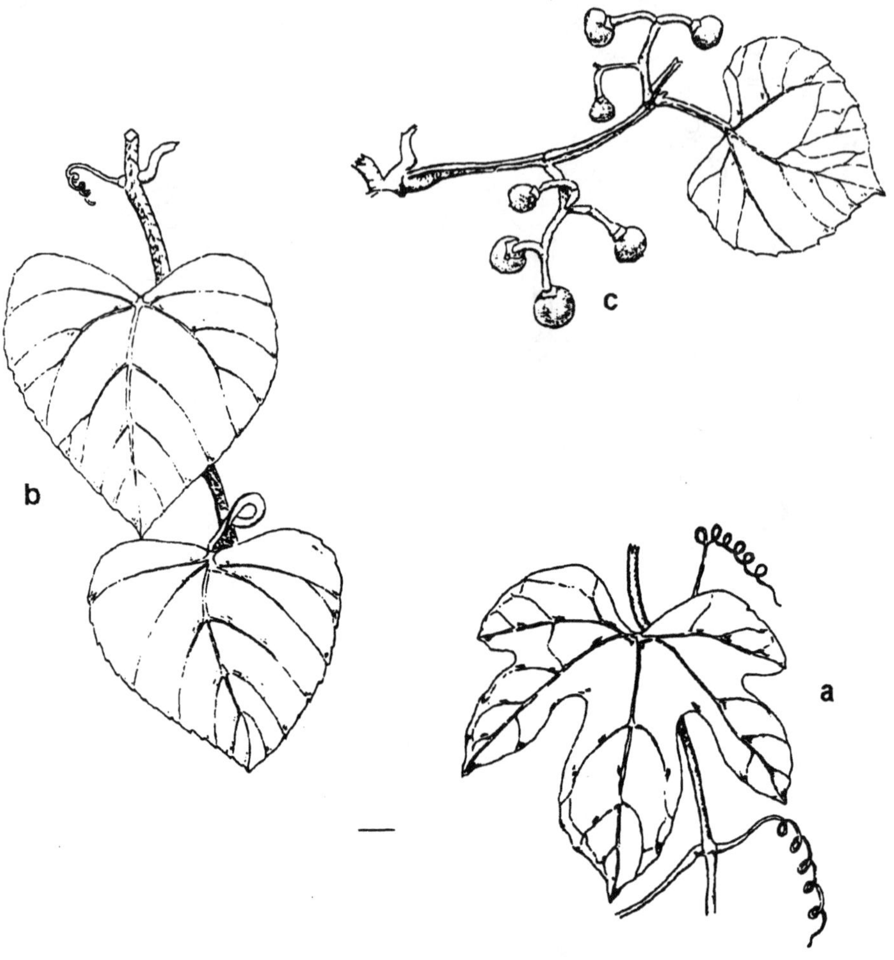

Climber. Stems green, turning brown when old, square, with narrow wings and grooves. Juvenile leaves deeply 3- or 5-palmately lobed (a). Adult leaves simple, alternate, usually with a tendril opposite each petiole (b); blade mid- to dark green, slightly glossy above, glabrous below; base usually cordate, occasionally truncate; margin with a minute tooth at end of each vein. Inflorescence branched, replacing tendrils. Buds reddish. Flowers minute, green, tinged with red. Fruit green, spherical with remains of persistent 4-partite calyx, maturing black, juicy, containing 1 seed (c).

Frequent in strand vegetation, especially on W side, and also in savanna on ridge top.

India through Malaysia, Indonesia, New Guinea to N Australia.

HCFH 1036/13568, 1740/13672, 2044/13799; Kuduk & Naoni UPNG 6115.

ZYGOPHYLLACEAE

Tribulus cistoides L.

(Henty & Pritchard: 167)

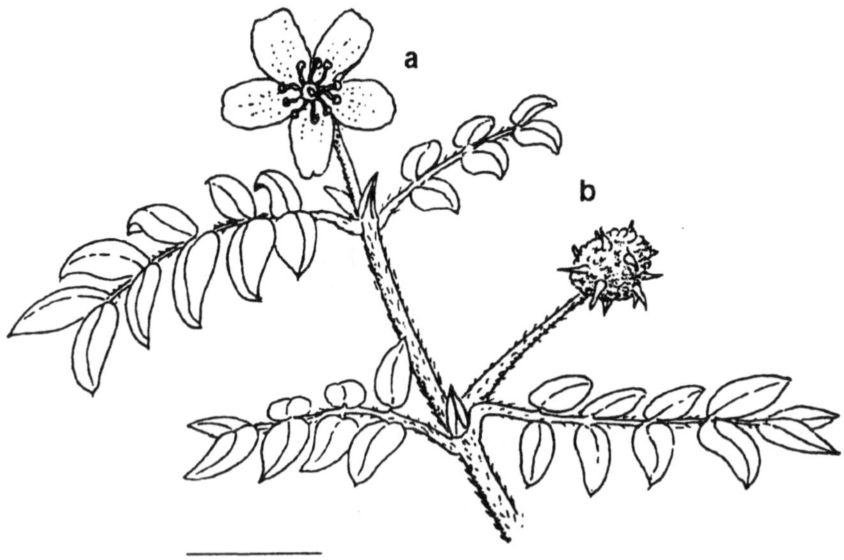

Prostrate herb with silvery-hairy pinnate leaves in unequal pairs. Flowers axillary, solitary, petals yellow, stamens 10. Fruit spiny, breaking into 5 portions.

Occasional, in mown grass around station buildings.

Pantropical weed.

JM 94/105.

MONOCOTYLEDONS

COMMELINACEAE

Commelina ensifolia R.Br. [not illustrated]

C. paleata Hassk

(Henty & Pritchard: 28, as *C. paleata*)

Decumbent herb. Leaves simple, alternate, blade narrowly ovate with sheathing leafbase; venation parallel. Flowers at ends of stems within a folded bract; petals 3, blue; anthers yellow, prominent.

Occasional in savanna on ridge top, along nature trail and as a weed in Station area.

India to New Guinea and Australia.

Frodin et al. UPNG 5238, 5255, 6664; HCFH 631/13162, 2012/13764; Tippet UPNG 824.

CYPERACEAE

Bulbostylis barbata (Rottb.) Clarke [not illustrated]

(Fl. Mal. 7: 539)

Herb to 10 cm high. Common in open areas and savanna. Old World and S USA.

Leach UPNG 7586.

Cyperus javanicus Houtt. [not illustrated]

(Fl. Mal. 7: 635)

Clumped herb to 1 m tall. On E side behind mangroves on white sand. Tropical Africa to Pacific islands and tropical Australia.

Kembi s.n. 22/7/75.

Cyperus rotundus L. [not illustrated]

(Fl. Mal 7: 604; Henty & Pritchard: 34)

Herb to 40 cm; spikelets red-brown. Disturbed ground near excavation site, Station area. Pantropical.

Frodin & Pernetta UPNG 6670; Kembi s.n. 22/7/75.

Sclerica brownii Kunth. [not illustrated]

(Fl. Mal. 7: 731)

Herb to 50 cm high. Savanna on ridge top. Tropical Australia, New Guinea, New Caledonia and Tonga.

Leach UPNG 7585.

Dioscorea bulbifera L.

Potato yam.

(Fl. Mal. **4**: 311)

Climber in grassland. Stems terete, unarmed. Leaves simple, alternate, cordate, glabrous. Fruit a papery capsule with 3 wings.

In savanna on ridge at N end.

W Africa through southern Asia to the Pacific Islands and Australia.

HCFH 2043/13798.

Dioscorea esculenta (Lour.) Burk.

Lesser yam.

(Fl. Mal. **4**: 307)

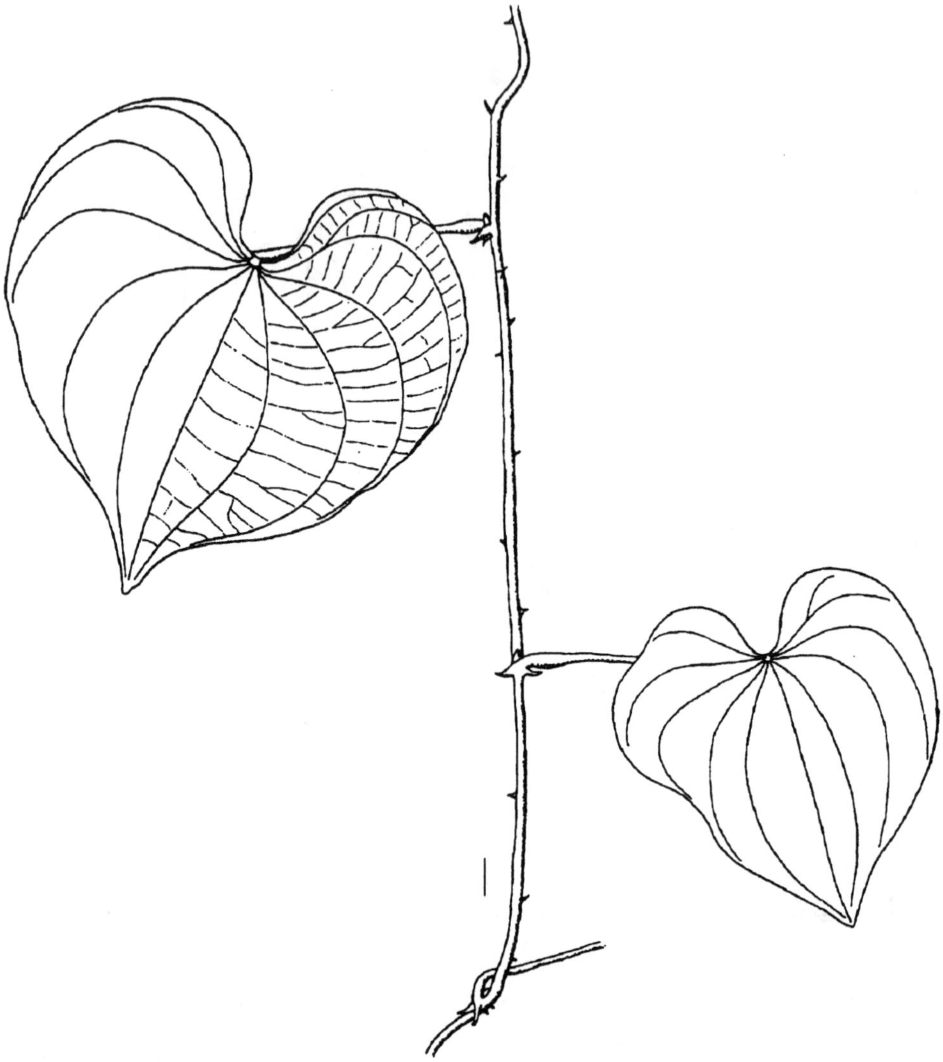

Climber up to canopy. Stems terete, green; prickles on stems and petioles, and 3 curved spines on stem at base of petiole. Leaves simple, alternate; blade, flat, plate-like, to 25-30 cm diameter, cordate; margin slightly sinuate. Flowers small, green, unisexual. Fruit a papery capsule with 3 wings.

In scrub on E side.

Wild in parts of Asia and perhaps New Guinea.

HCFH 2030/13782.

FLAGELLARIACEAE

Flagellaria indica L.

(Fl. Mal. **4**: 246)

Climber to c. 5 m high. Leaves simple, alternate; blade linear with tip modified as a tendril, dull olive-green, glabrous, base sheathing the stem. Flowers in dense terminal panicle, small, cream-yellow. Fruits globose, fleshy, orange-pink (a).

Frequent in scrub, just above strand line, E and W sides.

Africa, Indo-Malesia, to Polynesia and N Australia.

HCFH 922/13453, 1095/13627.

GRAMINEAE

Chloris barbata Sw. [not illustrated]

(Henty: 50)

Station area. Widespread throughout the tropics.

Frodin & Pernetta UPNG 6660; Lawong s.n. 24/8/88.

Cymbopogon sp. [not illustrated]

(Henty: 56)

SE side of island.

Kembi s.n. 22/7/75.

Cynodon dactylon (L.) Pers. [not illustrated]

(Henty: 57)

Station area. Tropical and warm temperate regions throughout the world.

Frodin & Pernetta UPNG 6652.

Dactyloctenium aegyptium (L.) Willd. [not illustrated]

(Henty: 60)

Station area. Widely distributed in tropical and warm temperate regions of the Old World; introduced into America.

Frodin et al. UPNG 5209, 6651, 6658; Lambinon 87/89; Lawong s.n. 24/8/88.

Digitaria longiflora (Retz.) Pers. [not illustrated]

(Henty: 77)

Station area. Old World tropics; introduced into America.

Frodin & Pernetta UPNG 6667; Lawong s.n. 24/8/88.

Eleusine indica (L.) Gaertn. [not illustrated]

(Henty: 88)

Station area. Pantropical.

Frodin & Pernetta UPNG 6647.

Eragrostis tenella (L.) Roem. & Schult. [not illustrated]

(Henty: 96)

Station area. Pantropical.

Frodin et al. UPNG 5239, 6656; Lambinon 87/94.

Heteropogon contortus (L.) Roem. & Schult. [not illustrated]

(Henty: 108)

Common on slopes. Tropical and warm temperate regions of the world.

Lepturus repens (G. Forster) R.Br. [not illustrated]

(Henty: 121)

Rocky seashore, W side. Widespread on sea-shores from E Africa to N Australia and Polynesia.

Kembi s.n. 22/7/75.

Oplismenus undulatifolius (Ard.) Roem. & Schult. [not illustrated]

(Henty: 133)

Scrub on E side and slope above shore. Warm temperate regions of N hemisphere, and Africa, India, Malesia to Australia.

Frodin UPNG 5247; HCFH 2008/13760.

Rhynchelytrum repens (Willd.) C.E. Hubb. [not illustrated]

(Henty: 166)

Widespread through savanna. Throughout Africa and introduced to most other tropical countries.

Frodin s.n. 1972; Lambinon 87/91.

Sorghum nitidum (Vahl.) Pers. [not illustrated]

(Henty: 176)

Savanna. S Asia to Australia.

HCFH 2017/13769.

Sporobolus diander (Retz.) P. Beauv. [not illustrated]

(Henty: 180)

Station area. India and Sri Lanka, Burma, Thailand, Malesia to Australia.

Frodin & Pernetta UPNG 6663.

Themeda novoguineensis (Reeder) Jansen [not illustrated]

(Henty: 185)

Savanna. Sulawesi and New Guinea.

Themeda triandra Forsk. [not illustrated]

Themeda australis (R. Br.) Stapf

(Henty: 184)

Savanna. Tropical and subtropical regions of the Old World.

Lambinon 87/98.

Dendrobium discolor Lindl.

(Millar: 18-21)

Woody herb to 80 cm high. Stems yellow, cane-like, covered by sheathing leaf bases. Leaves elliptic, thick, dull olive-green. Flowers in long axillary racemes, yellow, spotted brown; tepals twisted and crinkled (a). Fruit a green, ovoid capsule turning reddish when mature, to 2 cm diameter (b).

Rare on Motupore Island, only known wild from plants growing on a steep west-facing cliff but several similar plants are established on trees in Station area.

S New Guinea and N Australia.

Dipodium punctatum (Sm.) R. Br. [not illustrated]

(Millar: 72)

Terrestrial orchid that lacks leaves. Erect stems 30-40 cm high bearing up to 50 flowers, appearing in the wet season. Flowers pale pink to dark purple, spotted.

In grass on E side near nature trail.

N Australia, New Guinea, Vanuatu and New Caledonia.

HCFH 2045/13800.

PANDANACEAE

Pandanus tectorius Soland. ex Balf. f.

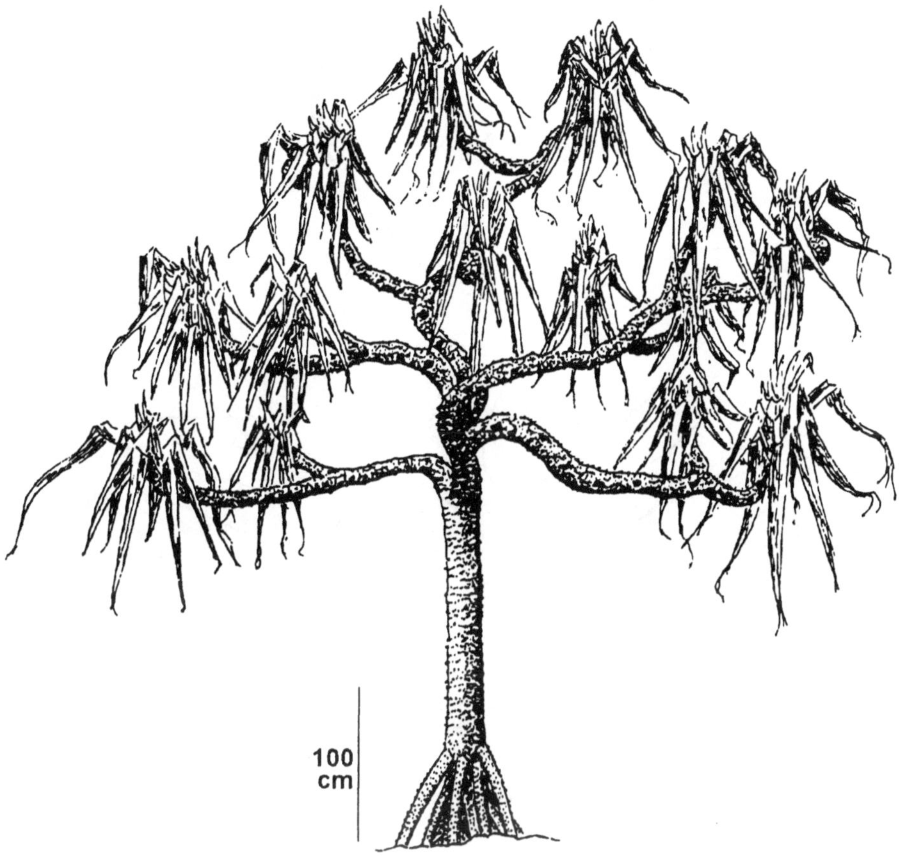

Sparsely branched tree. Stems covered with sharp prickles and leaf scars; trunk supported by stilt roots. Leaves simple, alternate, sword-shaped, to 1.5 m long, stiff, folded, with prickles along margin and mid-rib on lower side. Dioecious. Male flowers tiny, white, in large hanging clusters, surrounded by pale bracts. Female flowers and fruits aggregated into large, woody, globose balls.

Occasional in savanna and along strand line.

Tropical Pacific.

Smilax sp.

Woody scrambler. Stems green with occasional prickles. Leaves simple, alternate, mid-green, glabrous, with hard, downward pointing tip; 5 large veins arising at junction of blade and petiole; 2 tendrils at each node where a side shoot originates. Dioecious. Flowers in globular, axillary umbels, on side shoots, small, green-white. Fruit a spherical berry (a), juicy, green ripening red-black, containing 1 (-2) seeds (in section, b).

In scrub on E side of island behind Caretaker's house and scrub margin.

Frodin UPNG 5219; HCFH 1052/13584, 2029/13781. -

This may be *Smilax zeylanica* L. which Frodin (no date) records from Motupore and which occurs in S Asia and Malesia.

7. GLOSSARY

This glossary assumes a basic knowledge of the parts of the flower. The sources used for compilation include Henderson & Henderson (1960), Leach & Osborne (1985), Lincoln *et al.* (1982), Radford (1986), and Verdcourt (1979), and the diagrams are modified from Leach & Osborne (1985).

aculeate: having prickles or sharp points. Fig. 6.
acute: ending in a sharp point at an angle <45° . Fig. 7.
adpressed: flattened against the surface, usually referring to hairs; also spelt appressed.
adventitious: found in an unusual place; used especially of roots.
adventive: not native; transported into a new habitat by natural means or by human agency.
alternate: leaves, branches etc. placed singly at different positions along the axis; commonly used to include spiral arrangement. Fig. 8.
anastomosis: the joining of nerves and veins.
androecium: the male parts of a flower.
androgynophore: a stalk supporting the androecium and gynoecium.
annular: ring-shaped.
apiculate: ending in a short, sharp point.
aril: a fleshy outgrowth covering part of the seed.
article: a joint; the single-seeded segments of some leguminous fruits.
attenuate: gradually tapering at the base. Fig. 7.
axil: the upper angle between a leaf and the stem. Fig. 5.
axillary: occurring in the axil of a leaf or bract.

berry: a simple fleshy fruit, usually with several seeds.
bifid: split deeply into two parts; forked.
bipinnate: a compound leaf in which the primary divisions are themselves pinnate. Fig. 5.
bract: a modified leaf subtending an axis such as a flower or inflorescence.
bracteole: a small bract inserted on the pedicel and, in legumes, often paired below the calyx.

caducous: falling off at an early stage.
capitulum: a dense head-like inflorescence usually of sessile flowers. Fig. 9.
capsule: a dry, dehiscent fruit opening by valves or pores, with many seeds.
circumtropical: occurring throughout the tropics; = pantropical.
condensed cyme: a cymose inflorescence in which the flowers are short-stalked or sessile, and closely crowded together.
conduplicate: folded together. Fig. 6.
cone: a series of spirally arranged, often woody scales or sporophylls.
cordate: heart-shaped; a rounded leaf-base that is deeply notched. Fig. 7.
corona: an outgrowth of the corolla, often cup-shaped.
corrugated: deeply wrinkled or ridged.
crenate: margin notched with shallow, rounded teeth. Fig. 6.
cuneate: wedge-shaped at the base. Fig. 7.
cyathium: a cup-shaped involucre with stamens and a stalked gynoecium, each stamen and the gynoecium representing separate flowers; the inflorescence of *Euphorbia* spp.
cyme: an inflorescence in which the primary axis ends in a flower which opens first. Fig. 9.

deciduous: falling off at maturity; often referring to loss of leaves e.g. in the dry season.
decumbent: lying or trailing on the ground with the tip ascending.

decussate: in opposite pairs, but successive pairs oriented at right-angles to each other. Fig. 8.
dehiscent: spontaneously opening along particular lines (sutures) to release something, usually fruits releasing seeds or anthers releasing pollen grains.
dentate: with sharp teeth. Fig. 6.
dichotomous: dividing into 2 equal parts.
dioecious: having flowers of different sexes on separate plants.
disc: fleshy, sometimes nectar-secreting portion of the receptacle.
distichous: 2-ranked; arranged in 2 diametrically opposite rows; usually associated with alternate leaves. Fig. 8.
domatium, domatia (pl.): a small cavity, often inhabited by mites.
drupe: a simple fruit, usually with a fleshy mesocarp and a single seed surrounded by a hard endocarp.

ellipsoid: a solid object with an elliptic section or outline.
elliptic(al): oval-shaped. Fig. 5.
emarginate: notched at the apex. Fig. 7.
endocarp: inner layer of the pericarp or fruit wall, next to the seed; usually hard in drupes.
entire: with a smooth, even margin; not toothed, cut or divided. Fig. 6.
ephemeral: short-lived, transient; a plant that grows, reproduces and dies within a few days, or a flower that lasts for a day or less.
epicalyx: an aggregation of bracts just below the calyx, appearing like an extra calyx.
epiphyte: a plant growing on another plant for support or anchorage, but not parasitic.
erose: irregularly toothed or dentate, appearing as if gnawed. Fig. 6.
exfoliating: coming off in scales or layers.
exocarp: outer layer of the pericarp; = epicarp.
exserted: extending beyond another organ.

falcate: strongly curved; sickle-shaped. Fig. 5.
fasciculate: clustered in a small bundle or tuft.
felty: covered with matted hairs.
ferrugineous: with the appearance of rusty iron.
fissure: a deep groove or furrow, usually referring to bark.
flaccid: loose, limp, drooping.
flange: a flat projection.
follicle: a simple, dry, dehiscent fruit, formed from a single carpel, which opens by one suture only.
frond: a leaf, especially of a fern or palm.

glabrescent: becoming glabrous or devoid of covering at maturity.
glabrous: without hairs or scales; smooth.
glaucous: with a bluish-green, waxy covering on the surface.
gynoecium: the female parts of a flower.
gynophore: a stalk supporting the gynoecium.

habitat: locality, site and particular type of local environment occupied by an organism.
hair: a thread-like outgrowth of the epidermis consisting of one or more cells.
haustorium, haustoria (pl.): outgrowth of a parasitic plant that draws food from the host plant; sucker.
herbaceous: non-woody.
hermaphrodite: an organism with both male and female reproductive organs in one individual; a flower with both an androecium and a gynoecium.
hispid: coarsely and stiffly hairy.
hoary: covered with short, white hairs.
hypocotyl: the portion of a stem below the cotyledons in the embryo and seedling.

imparipinnate: pinnate with an odd number of leaflets, ending in a terminal leaflet. Fig. 5.
indumentum: covering of hairs or scales.
introgression: the spread of genes of one species into the gene pool of another by hybridization and back-crossing.
involucral: forming an involucre.

involucre: bracts in a whorl, forming a calyx-like structure round the base of an inflorescence, such as a capitulum or umbel.
involute: with the margins rolled inwards. Fig. 6.
isobilateral: a form of symmetry where a structure is bilaterally symmetrical in two planes at right-angles to each other.

keel: the lower or anterior petal(s), shaped like the keel of a boat.

lamina: blade of a petal or a leaf; a thin, flat piece of tissue.
lanceolate: lance-shaped, i.e. broadest towards the base, narrow and gradually acute at the apex. Fig. 5.
latex: milky sap or juice.
lax: loose; not dense.
LBR: length/breadth ratio.
leaflet: individual unit of a compound leaf.
lenticel: a ventilating pore in an angiosperm stem or root, often corky.
liane, liana: climbing plant with a woody stem.
linear: long and narrow with parallel margins. Fig. 5.
linear-lanceolate: narrow with almost parallel sides, between linear and lanceolate in shape.
lithophytic: growing on rocky or stony ground.

megasporophyll: a modified leaf bearing ovules or megasporangia.
mericarp: a one-seeded indehiscent part of a schizocarp, which is a dry fruit that splits into two or more one-seeded parts i.e. splitting along the junction between the carpels and not along a suture within a single carpel.
mesocarp: the middle layer of the fruit wall or pericarp.
microsporophyll: a modified leaf bearing pollen sacs or microsporangia.
moniliform: arranged like a chain of beads, constricted at regular intervals.
monoecious: having unisexual flowers but both sexes on the same plant.
mucronate: ending abruptly in a short, stiff point. Fig. 7.
mucronulate: diminutive form of mucronate, i.e. tipped with a small mucro or point.

node: the point on a stem at which one or more leaves are attached; a point of branching of a stem. Fig. 5.

ob-: used in front of the description of a shape to indicate a reversal of shape e.g. obovate.
oblanceolate: inversely lanceolate. Fig. 5.
oblate: a flattened sphere or circle.
oblong: longer than broad with parallel sides. Fig. 5.
obovate: inversely egg-shaped; a leaf with the narrow end attached to the stem. Fig. 5.
obreniform: kidney-shaped, with the broadest part nearest to the stem.
opposite: two organs, such as leaves, arising at the same level on opposite sides of the stem. Fig. 8.
ovate: egg-shaped, the broad part being nearer to the stem. Fig. 5.
ovoid: somewhat egg-shaped; a solid figure with an ovate cross-section.
ovule: structure containing the egg and developing into a seed after fertilization.

pagoda: a building shaped like a tower of truncated pyramids.
palaeotropical: occurring in the Old World tropics, i.e. Africa, Indo-Malesia and Australasia.
palmate: divided into segments like the fingers of a hand. Fig. 5.
palmately nerved: the main veins of a leaf or midribs of leaflets meeting at one point
palmatifid: a leaf palmately cut into lobes connected by lamina as well as midrib. Fig. 5.
panicle: a branched racemose inflorescence. Fig. 9.
pantropical: occurring throughout the tropics; = circumtropical.
paripinnate: pinnate with an even number of leaflets and no terminal leaflet. Fig. 5.
pedate, pedately compound: a leaf divided into 3, then the lateral parts each divided again.
pedicel: the stalk of a single flower.
pedicellate: stalked (flower).
peduncle: the stalk of an inflorescence or part of an inflorescence.
pedunculate: stalked (inflorescence).
pendulous: hanging; bending downwards.
pentamerous: composed of 5 parts; in whorls or multiples of 5.

pericarp: fruit wall, including the epicarp, mesocarp and endocarp.
petiolate: stalked (leaf); having a petiole or leaf-stalk.
petiole: the stalk of a leaf. Fig. 5.
phyllode: an enlarged, flattened petiole which resembles and functions as a leaf.
pinna, pinnae (pl.): the divisions of a pinnate leaf (i.e. = leaflets), or the divisions of a bipinnate leaf.
pinnate: a leaf composed of 2 or more leaflets arranged along each side of a common stalk or rachis.
pneumatophore: erect roots that are upward extensions of subterranean root system, adapted for gaseous exchange.
prickle: a sharp outgrowth from the epidermis.
propagule: any part of an organism that is capable of giving rise to a new individual.
prostrate: lying along the ground.
pseudo-: false; hence pseudocyme and pseudoraceme.
puberulous: sparsely covered with down or fine hair.
pubescence: a downy or hairy covering.
pubescent: covered with short, soft hairs.

raceme: an elongated, unbranched racemose inflorescence in which the flowers are borne on pedicels. Fig. 8.
racemose: an inflorescence whose growing point is indeterminate, i.e. usually continues to add to the inflorescence, and in which there is usually no terminal flower; the youngest flowers are towards the apex.
rachis: the axis of an inflorescence or of a compound leaf above the petiole. Also spelt rhachis.
recurved: curved downwards.
reflexed: bent sharply backwards.
reticulate: a network, usually of veins, formed by repeated branching and anastomosis.
revolute: the edges rolled downwards or backwards. Fig. 6.
rhizome: a modified, underground stem, usually growing horizontally; the stem of a fern, both aerial and underground.
rhombic: shaped like the diamond on a playing card.
rhombic-elliptic: shape between rhombic and elliptical.

salviform: trumpet-shaped; a corolla with a slender tube and the apical part nearly at right-angles to it.
scale: a small, flat structure.
scandent: climbing or trailing over other vegetation, usually without special climbing organs.
scarious: thin, dry, membraneous.
schizocarp: see **mericarp**
scrub: a community dominated by shrubs.
scurfy: scaly epidermal covering.
seed bank: store of dormant seeds buried in soil.
serrate: toothed or notched, the teeth pointing forward like those of a saw. Fig. 6.
serrulate: finely toothed or notched. Fig. 6.
sessile: without a stalk.
simple: composed of one piece; not compound.
sinuate: having a wavy outline or margin, in the same plane as the organ. Fig. 6.
sorus, sori (pl.): a group of fern sporangia, usually on the underside of a frond.
spike: an unbranched, elongated, racemose inflorescence with sessile flowers. Fig. 9.
spikelet: a small spike; the basic inflorescence unit in Gramineae and Cyperaceae.
spine: a sharp-pointed process derived from subepidermal tissue.
spinose: bearing many spines.
standard: the rear or upper petal, e.g. in a papilionoid flower.
stellate: star-shaped.
stilt root: branched, arched, aerial roots that arise from the trunk and lower branches.
stipular: like stipules or growing in places of stipules, e.g. thorns.
stipule: one of two small leaf-like or membraneous structures near the base of the petiole. Fig. 5.
stone: the hard endocarp of a drupe, containing the seed.
succulent: thick and fleshy, juicy.
superior: situated above another organ or part.
suture: the line of dehiscence of a carpel.

syncarp: an aggregated fruit with united carpels.

tendril: a climbing organ, often slender and curling, formed from the whole or part of a stem, leaf, or petiole.
tepal: part of a perianth not differentiated into petals and sepals.
terete: cylindrical, circular in transverse section; not ridged.
tesselated: chequered; markings or colours arranged in squares.
thorn: a sharp-pointed woody structure formed from a modified branch.
trifoliate: a compound leaf consisting of 3 leaflets; = ternate. Fig. 5.
trullate: shaped like a bricklayer's trowel i.e. broadest below the middle with 2 equal straight sides meeting at the apex and 2 shorter sides meeting at the base. Fig. 5.

umbel: an inflorescence in which the pedicels all arise from the top of the main stem and all the flowers are in ± the same plane. Fig. 9.
umbellate: arranged in umbels.
undulate: corrugated or wavy in a plane at right-angles to the surface. Fig. 6.
unifoliate: with one leaf.
unifoliolate: with one leaflet.

valve: the pieces of the pericarp of a dehiscent fruit; a door or flap-like structure.
velutinous: velvety; covered in very fine, dense, short, upright hairs.
verticillaster: a much condensed cyme with the appearance of a whorl, but actually arising in the axils of opposite leaves.
verticillate: whorled.
viviparous: germinating while still attached to the parent plant.

wing (petal): one of two lateral petals in a papilionoid flower.

zygomorphic: having only one plane of symmetry; bilaterally symmetrical.

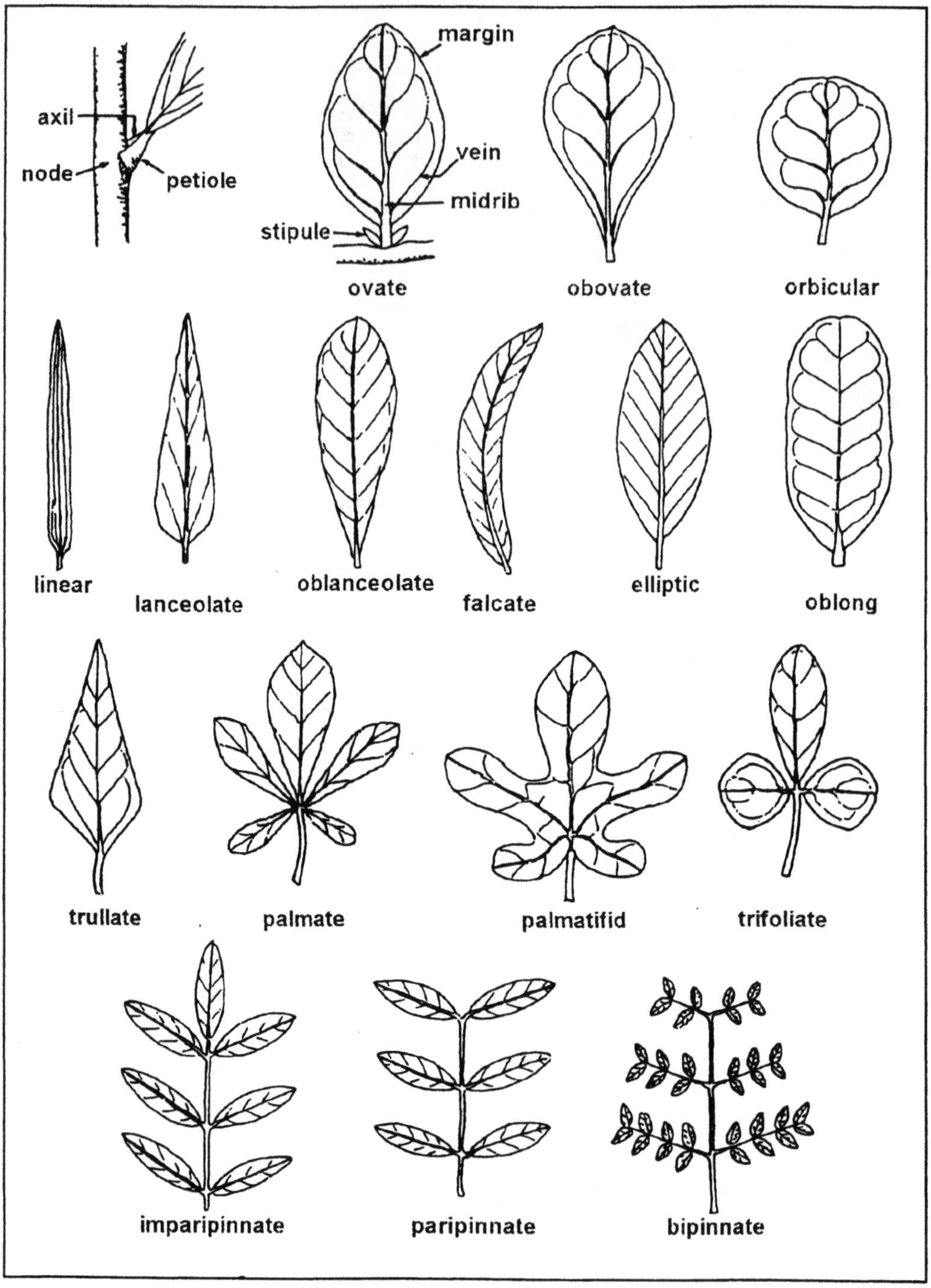

Figure 5. Leaf shapes and structures.

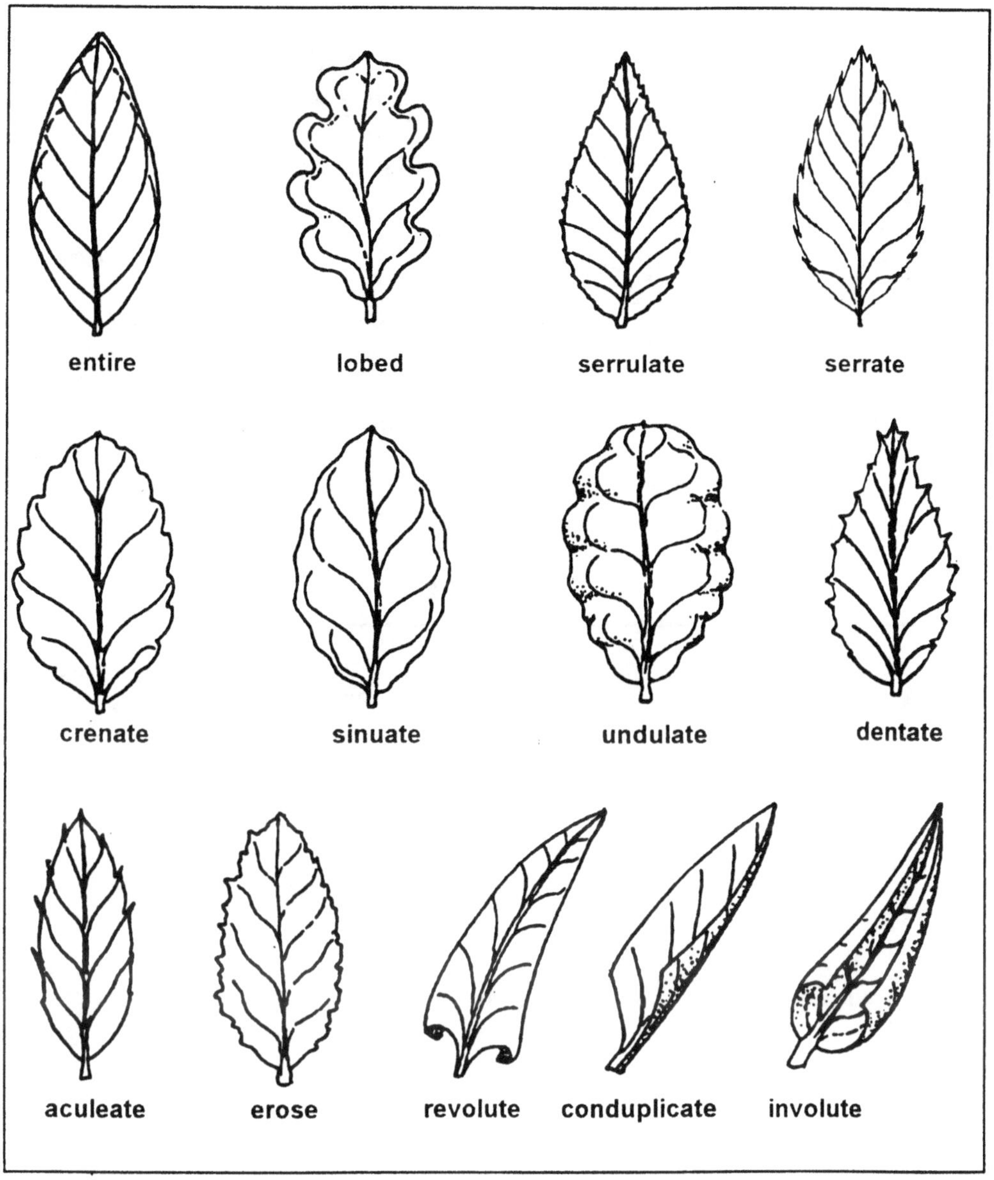

Figure 6. Leaf margins.

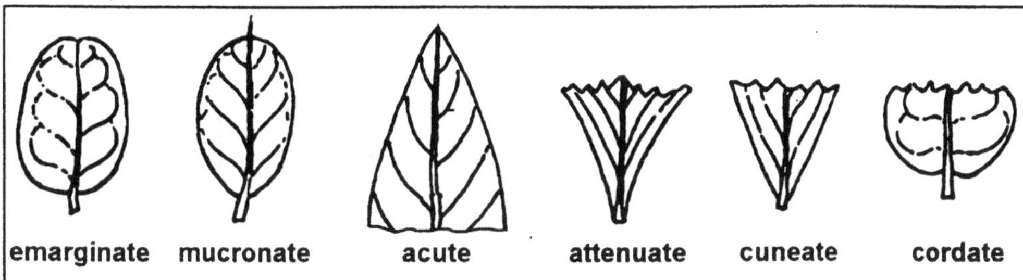

Figure 7. Leaf apices and bases.

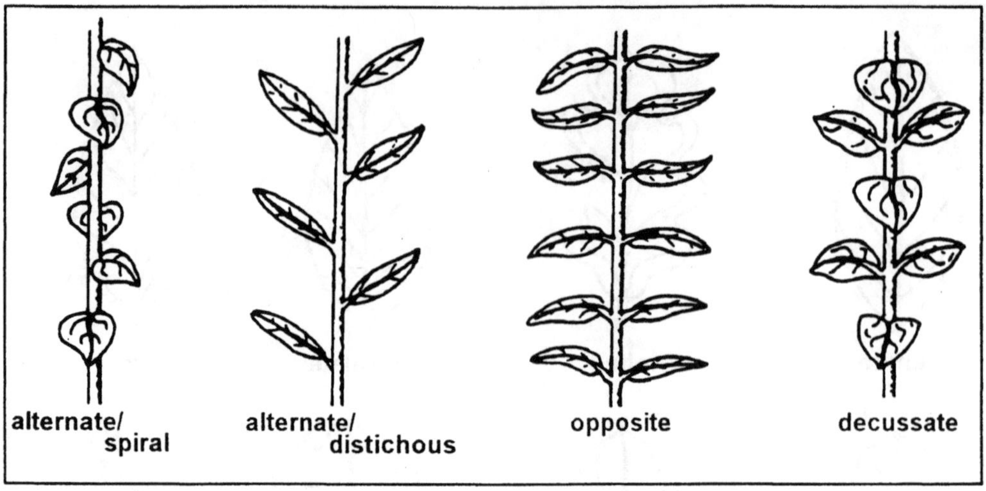

Figure 8. Leaf arrangement.

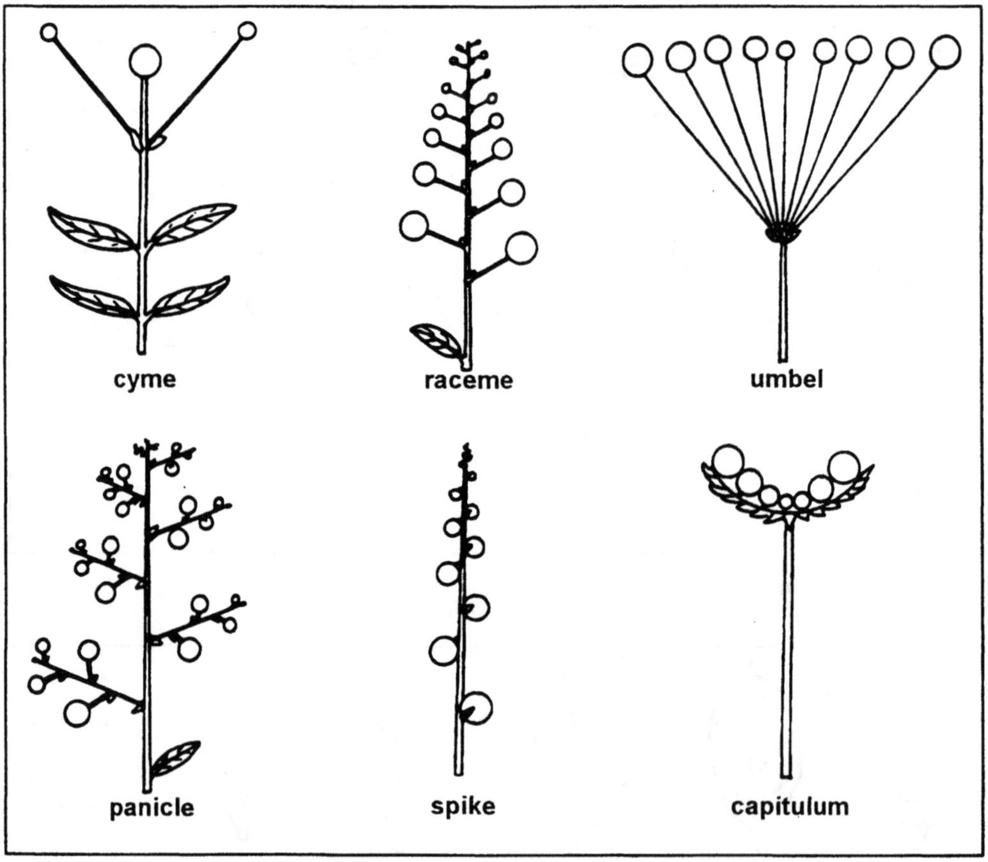

Figure 9. Inflorescence types.

8. BIBLIOGRAPHY

Copies of all unpublished documents are available in the files of Motupore Island Research Department, UPNG.

Barratt, C.J. (1972). Untitled report on vegetation plots on Motupore Island. Unpublished manuscript.

Brock, J. (1988). *Top End native plants.* John Brock, Darwin.

Brooks, K., Chambers, M., Frodin, D., Haig, D., Humphreys, G., Kidd, R. & Polunin, N. (no date). *The Motupore Island nature trail.* Unpublished manuscript.

Doutch, H.F. (1972). Palaeogeography of northern Australia and New Guinea. Pp. 1-10 in D. Walker (ed.) *Bridge and barrier, natural and cultural history of Torres Strait.* Australian National University, Research School of Pacific Studies, publication BG/3.

Eden, M.J. (1974). The origin and status of savanna and grassland in southern Papua. *Institute of British Geography, Transactions* **63**, 97-110.

Frodin, D.G. (no date). Unpublished checklists of the plants of Motupore island.

Frodin, D.G., Huxley, C.R. & Kirina, K.W. (1974). *Mangroves of the Port Moresby Region.* University of Papua New Guinea, Dept. of Biology Occasional Paper No. 3. (revised version by D.G. Frodin & G.J. Leach, 1982).

Frodin, D.G. & Johnstone, I.M. (no date). *A self-guiding tour of the plants of Motupore Island.* (unpublished manuscript).

Gunn, C.R. & Dennis, J.V. (1976). *World guide to tropical drift seeds and fruits.* Demeter Press, New York.

Henderson, I.F. & Henderson, W.D. (1960). *A dictionary of scientific terms.* (7th edition by J. H. Kenneth). D. van Nostrand, New York.

Henty, E.E. (1969). *A manual of the grasses of New Guinea.* Botany Bulletin No. 1. Department of Forests, Lae.

Henty, E.E. & Pritchard, G.H. (1982). *Weeds of New Guinea and their control.* Botany Bulletin No. 7. Department of Forests, Lae.

Heyligers, P.C. (1965). Vegetation and ecology of the Port Moresby-Kairuku area. Pp. 146-173 in J.A. Mabbutt *et al.* (eds). *Lands of the Port Moresby-Kairuku area, Papua-New Guinea.* Land Resources Series 14. CSIRO, Melbourne.

Heyligers, P.C. (1966). Observations on *Themeda australis-Eucalyptus* savanna in Papua. *Pacific Science* **20**, 477-489.

Heyligers, P.C. (1972). Analysis of the plant geography of the semi-deciduous scrub and forest and the Eucalypt savanna near Port Moresby. *Pacific Science* **26**, 229-241.

Heyligers, P.C. (1982). Semi-deciduous scrub and forest and Eucalypt woodland in the Port Moresby area. Pp. 437-457 in J.L. Gressitt (ed.) *Biogeography and ecology of New Guinea. Monographiae Biologicae* **42**. Dr. W. Junk, The Hague.

Heywood, V.H. (ed.) (1978). *Flowering plants of the world.* Oxford University Press, Oxford.

Leach, G.L. & Osborne, P.L. (1985). *Freshwater plants of Papua New Guinea.* University of Papua New Guinea Press, Port Moresby.

Lincoln, R.J., Boxshall, G.A. & Clark, P.F. (1982). *A dictionary of ecology, evolution and systematics.* Cambridge University Press, Cambridge.

Mabberley, D.J. (1987). *The Plant Book.* Cambridge University Press, Cambridge.

McGregor, K.R. & Huber, M.E. (1993). A bibliography of research relating to Motupore Island, Papua New Guinea. *Science in New Guinea* **19**, 59-76.

Menzies, J.I. & Sengo, J. (1993). Towards a monsoon forest? Vegetation changes on Motupore Island. *Proceedings of the New Guinea Biological Society, September 2-3, 1993.* Universitas Cenderawasih, Manokwari, Indonesia.

Moore, N.H. (1982). *Oceanographic observations in the western sector of the Papuan coastal lagoon.* Final report of the Papuan coastal lagoon project, Motupore Island Research Department, University of Papua New Guinea, Port Moresby.

Peekel, P.G. (1984). *Flora of the Bismarck Archipelago for naturalists.* Office of Forests, Lae.

Pijl, L. van der (1957). The dispersal of plants by bats. *Acta Botanica Neerlandica* **6**, 291-315.

Pijl, L. van der (1982). *Principles of dispersal in higher plants* (3rd ed.). Springer-Verlag, Berlin.

Radford, A.E. (1986). *Fundamentals of plant systematics.* Harper & Row, New York.

Royen, P. van. (1963). *Vegetation of the island of New Guinea.* Department of Forests, Lae.

Swadling, P. (1981). The settlement history of the Motu and Koita speaking people of the Central Province, Papua New Guinea. Pp. 239-251 in D. Denoon & R. Lacey (eds) *Oral tradition in Melanesia.* University of Papua New Guinea and Institute of Papua New Guinea Studies, Port Moresby.

Tomlinson, P.B. (1986). *The botany of mangroves.* Cambridge University Press, Cambridge.

Verdcourt, B. (1979). *A manual of New Guinea legumes.* Botany Bulletin No. 11. Office of Forests, Lae.

Worthing, M. (1978). *Raw material sources for potsherds from Motupore Island midden, Port Moresby, Papua New Guinea.* Unpublished manuscript.

9. INDEX TO PLANT NAMES

The main references to the accounts in chapter 5 are given in **bold**. Page references to names in keys are omitted.